85467022X

D1743485

Peter Gibson.

LIGHTING CRAFTS

GLEN POWNALL'S CREATIVE LEISURE SERIES **Lighting Crafts**

SEVEN SEAS PUBLISHING PTY LIMITED *Wellington and Sydney*

Acknowledgements

The publisher and author wish to thank
Floral Boutique, Guy Ngan and Pablo Industries Ltd.
for their assistance.

All photographs taken by Richard Silcock.

ISBN 85467 022 X

© Copyright 1974 Glen Pownall
Wellington, New Zealand
Printed by Dai Nippon Printing Co. (Hong Kong) Ltd

Contents

Lighting Crafts

INTRODUCTION
Painting in light, sculpturing in colour is now a reality, a possibility for every home. Artificial light on the domestic scene is undergoing its second great revolution.

The first major change in centuries in artificial lighting occurred less than 150 years ago, a fact not always realised. Fifteen thousand years ago, in the palaeolithic period, there were artists living in the Lascaux caves of the Dordogne area of France, who painted animal pictures. They used a type of grease lamp held in a hollow stone and some of these lamps have been found with discernible traces of animal fat still able to be identified. By the early 19th century AD, the great majority of the world's people were using lights. Lights which had shown no significant improvement, and gave no more illumination than the lamps of those Lascaux cave dwellers.

Thus, for at least a known fifteen thousand years there was virtually no technological progress in lighting craft. The world of light first began to change between 1825 and 1850 with improvements in candlemaking materials. The candles offered in a convenient and workable form then, for the first time ever, a way of prolonging a working day beyond the hours of daylight. Adequate and reliable night illumination was one of the truly great steps forward in man's progress. Good lighting had an impact on people of the mid 19th century which is often disregarded by social historians. Now, some one hundred years plus beyond that time, artificial lighting is several times more effective than the best of wax candles.

It is not the purpose of this book to consider the fantastic rate of progress in technology over the past hundred years. The text is concerned rather with a new lighting revolution, the effects of which we are only just becoming aware. Light is coming to be appreciated as more than just a means of seeing, it is being recognised as a source of sense stimulation, a highly sophisticated means of creating an emotional atmosphere. It affords a way of communication between people; an artistic medium, through which the emotions of the artist can be transmitted to an audience.

This book is also concerned with, and considers in depth the very live subject of light as a means of personal impression. The sole criterion of 'good lighting' within these pages is a judgment based on the emotional impact of a properly designed and executed lighting effect. Primitive as well as complex lighting sources are examined and explained to the amateur. This has to be a 'how-to' book, for the flexibility of the medium is so great; the philosophy of light, as an art medium in its own right, so new, that there are no real professionals in the field. Everyone engaged in lighting craft is an experimenter, an entrepreneur who cannot be instructed, only offered some guidance as to the path to follow. The advice given is practical and down to earth allowing craftsmen to build for themselves a very wide range of creative lighting objects.

6

By candlelight

Candlelight is romantic. If this were a different kind of book, consideration could be given to candlelight in the romantic setting of crisp table linen, sparkling crystal, polished silver, savoury meats, good wines, an ardent swain, and the effect of all this glamour on feminine emotions. However, it is not that sort of text.

Tallow candles

Animal fats, of which the semi-purified form is tallow, do not make good candles, yet, with the exception of bees-wax, which was in limited supply, tallow was the only readily available material for candle making until well into the 19th century. As late as 1838, Count Rumford, (an inveterate investigator of any phenomena which caught his attention) gave the following comparative figures:

'An ordinary tallow candle (well lighted) at the beginning gives a light equal to 100.' *Note* In 1838 there were no standard units of illumination and the figure 100 is merely Count Rumford's indication of the maximum brilliance of the particular candle he was examining. To continue: 'After 11 minutes burning, the figure was 39; in 19 minutes, 23; in 29 minutes, 16; upon being resnuffed it regained its former brilliance. A bees-wax candle continued to give equal brilliance as long as it burnt. In comparison a tallow candle consumed 101 parts by weight to the wax candle's 100 parts by weight when the tallow candle was newly resnuffed, but the consumption rose to 229 parts by weight when in need of resnuffing and the light was very dim.'

In its own way Count Rumford's report is a remarkable application of the scientific method and introduces a number of points which a candlemaker of today must understand. First comes the question of the greatly increased consumption of tallow with the very low light output. This was caused by deficiency in design of candles which is still prevalent today, in effect the wrong choice of wick. No wicking material then available prevented guttering tallow candles, see figure 1/1, the partly burnt wick bending over and forming a 'gutter' or 'gully' down which the molten tallow would run. The only remedy with low and variable melting-point tallows was to snip off the charred end of the wick every twenty minutes or so.

Relighting a candle with a tinder box, flint and steel had somewhat uncertain results and candle snuffing (without dousing the flame) was and still is an act requiring luck, judgment and dexterity. A tallow candle left unattended guttered badly, it being estimated by one of the savants of last century that less than one-twentieth of the tallow was burned, the rest running to waste.

A large wick would have ensured that the flame was adequately supplied with fuel and that the wick was fully consumed. An oversized wick, however, was almost certain to smoke through excess of hot fat in the flame. Partly

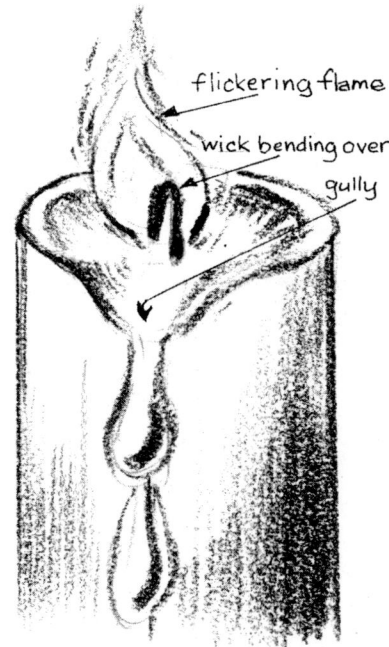

Figure 1/1

stone and shell lamps

ball of cotton
wicking burning
in coconut oil

coconut

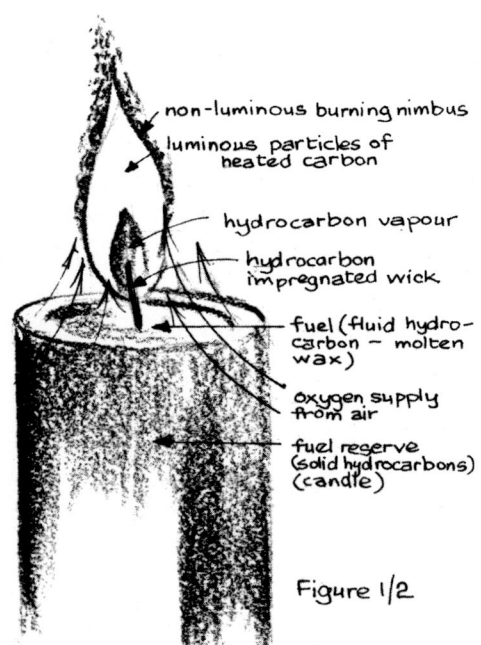

non-luminous burning nimbus

luminous particles of
heated carbon

hydrocarbon vapour

hydrocarbon
impregnated wick

fuel (fluid hydro-
carbon – molten
wax)

oxygen supply
from air

fuel reserve
(solid hydrocarbons)
(candle)

Figure 1/2

burnt animal fat gives off a vile odour, offensive to even our ancestors, who were not unduly sensitive about their personal hygiene.

It is interesting to note that 'snuffing' was not the act of putting out the light, for quite elaborate cutting instruments were used to cut (snuff) the wick and catch the hot charred excess wick while still leaving the candle burning. Extinguishers were small metal cones which were supposed to put the candle out and thus prevent the permeating stench of partly burnt tallow from escaping. In modern idiom candle extinguishers are wrongly offered as candle snuffers, a misunderstanding of the verb 'to snuff'.

The mechanics of a candle flame
Tallow candles have been mentioned here partly owing to the inherent interest of giving some not so well known facts about the difficulties suffered by our fairly immediate great-grandparents but also to justify a better understanding of correct design in candlemaking, a subject which is usually taken for granted. There are two choices that can be made in a book devoted to craftwork and which are particularly applicable to candlemaking. There is the usual, very simple method of advising the use of specific materials and techniques, plus recommending proprietary factory-made wicks. The alternative is to give facts about the mechanics of candle design and the properties of various materials, then allow craftsmen the creative satisfaction of adding to their own understanding in producing entirely personal and unique creations. It is this second approach that is the basic philosophy of all books in the Creative Leisure series and which is particularly accented in this volume.

For the reason just given it is now proposed to discuss some technical considerations. Consider figure 1/2 which shows just the head of a burning candle. Yet a candle flame is surprisingly complicated, supporting thereby those critics who claim that science is a means of making the easy sound difficult and the simple, complex.

The figure shows a number of technically important facts, as follows:
1. A candle operates through the chemical combination (burning) of hydrocarbons with oxygen. Hydrocarbons are a very large class of chemical compounds derived from once living matter (organic substances) and contain the elements carbon and hydrogen in various combinations.
2. Paraffin wax is a typical example of a hydrocarbon. To those interested in chemistry, paraffins have the general formula $C_nH_{2n} + 2$. When ignited in the presence of oxygen, there is an exothermic reaction (exothermic means giving off heat) between the hydrocarbon and the oxygen, resulting in the formation of water vapour and carbon dioxide when combustion is complete.
3. The success of the reaction is entirely dependent on the correct proportion of hydrocarbon vapour and oxygen.

All this may savour of dull theory but it has important practical applications in the function of a candle. For instance:
1. The air supply (which contains the oxygen) must be sufficient to combine fully with all the carbon in the hydrocarbon. The hydrogen content will combine (burn) first. If there is too much hydrocarbon or not enough air (the practical results of both are the same), the hydrogen will burn and free

8

Grecian style lamp c400 B.C.

carbon (soot) will be released. The flame will smoke.

2. The exact opposite occurs if there is too much air and not enough wax. The flame either becomes overcooled and goes out or the wick burns away, causing continuous flickering and a smell of burning wick material. A third possibility is the flame burning too close to the wax, overheating the sides of the candle so that the candle gutters.

An over supply of wax is caused either by a restricted air supply (as when the candle is placed at the bottom of a glass vessel) or through too large a wick. An over supply of air is normally caused by a wick which is too small or through using a material for it which burns away too fast. In a draught or breeze there will be a continual change between these two extremes and the candle will alternately smoke and gutter.

Whether by science or commonsense, if the just mentioned principles are observed, the candle as a source of light will be a success. Whether it will be a piece of decorative art work is another matter which will be considered in depth later.

Candles from nature
Possibly the greatest satisfaction in craft work comes from the direct use of nature's own materials, being absolutely independent of machines or machine-made products. Although tallow candles are most unsatisfactory (without extensive treatment of the raw material) there are other natural waxes which make really good candles.

Natural candle wicks
The ancient Chinese when making candles used as wicks spills of paper wrapped around a rush centre. It is not known by this writer what species of rush was used by the Chinese for this purpose but it is almost certain to have been a member of Family *Juncaceae* (the true rushes), for the species of this family have world wide distribution.

In Europe the soft rush, *Juncus effusus*, or less commonly *J. communis* (the common rush) were used for rush lights, which were produced as follows:
1. The rushes were gathered fully grown, but still green.
2. About 18 inches of the 'prime' (the centre of the rush) was left by cutting away both ends.
3. The green outer coating was stripped from each length, with the exception of a narrow strip. See figure 1/3.
4. If the rushes were not to be peeled immediately on cutting they were held under water until ready to be used.
5. The peeled rushes were then sun dried.
6. The better type of rush lights were dipped in tallow to which a modicum of bees-wax had been added. Continued dipping produced a candle.
7. The class of rush-light as used by the poorer country families required that the prepared rushes were placed in a shallow dish and any excess fat was poured over them.

A quality rush-light is said to have burned for about half an hour, and sixteen hundred rushes (1 lb) used 6 lb of tallow. Apart from the short burning time, rush-lights also suffered the disadvantage of leaving a line of tallow drops on the floor. It can be assumed that any rush can substitute as a

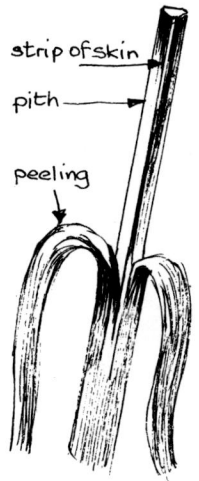

strip of skin

pith

peeling

Figure 1/3

Roman bronze lamp after Diderot

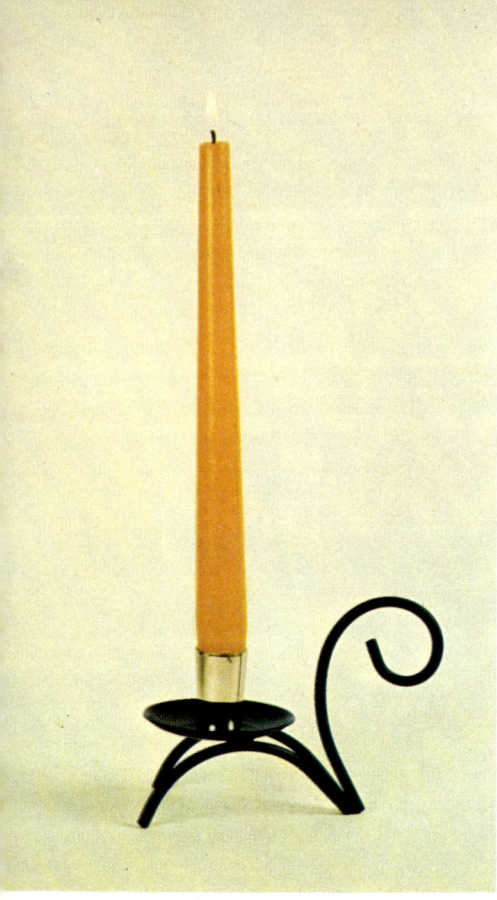

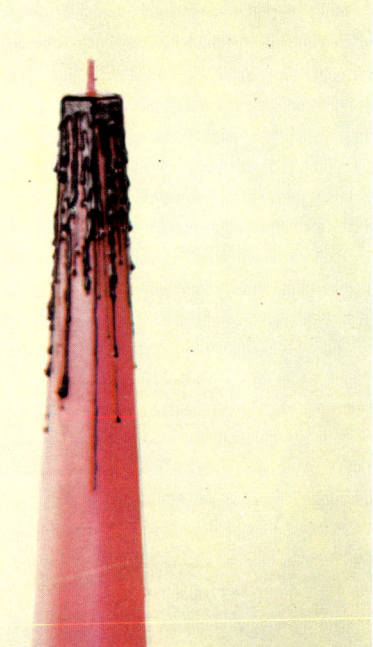

candle wick, particularly when treated in this way. The unpeeled strip of the skin causes the tip of the wick to bend sideways into the air and burn, thus preventing guttering.

Natural candle waxes

Bayberry wax
The finest of all natural waxes for candlemaking comes from the fruit of an ornamental tree which is a native of eastern North America. The bayberry *Myrica cerifera*, sometimes *M. carolinensis* produces fruits with a mealy crust of wax on the surface. The procedure of collecting and refining this wax is as follows:
1. Collect a considerable quantity of bayberry fruit.
2. Boil the fruit, strain through muslin while still very hot; fold the muslin so that the debris is inside it and reboil with the muslin in the water; restrain and skim the wax off the surface.
3. If necessary repeat this operation several times until the wax becomes clear, sub-transparent, pale green and fragrant.

The patience required to produce bayberry candles is not wasted as the best, most fragrant natural wax candles are made from this fruit. For the technically minded, bayberry bark wax can be obtained from the root bark of *Myrica cerifera* but extraction requires extended cooking in a pressure vessel. A household type pressure cooker achieves limited success but is worth trying. The resulting mixture of waxes and gum resins contains myricic acid, tannin, red coloured, astringent, acrid resin, gum and starch as well as palmitic, myristic and lauric acid esters. A candle made from this wax also acts as an insect repellent.

Chinese vegetable tallow
This is obtained from the fruit of a fairly large tree (*Sapium sebiferum*) endemic to China and Japan, and consists largely of palmitic acid and has a faint smell similar to violets. It was this product which enabled the Chinese to produce quality candles long before such luxuries were known in Europe.

Chinese wax
As distinct from vegetable tallow, Chinese wax is the exudation of two species of insects *Coccus ceriferus* and *C. peta* deposited on the branches of a certain species of ash tree in Western China. The wax is scraped from the tree, refined by boiling, then used in candlemaking. It consists chiefly of ceryl cerotate and has a comparatively high melting point, 92°C, thus making it a long burning candle.

Candililla wax
A plant (*Pendilanthus pavonis*), a native of northern Mexico and southern Texas is covered in a wax which coats it completely. With a melting point of 68° to 70°C this brownish yellow wax makes serviceable candles.

Bees-wax
The comb of the honey bee has never lost pride of place as the source of the best wax for candle making. Chemically it consists of about 80% myricin,

10

(ceryl myristate) with some cerotic acid esters and a low percentage of other matter. The melting point is 62° to 65°C, it has a honey-like odour and a slightly balsamic, pleasant flavour. Bees-wax candles are required in certain religious observances and bees-wax is a preferred ingredient in quality candles of all types. It is discussed more fully in Chapter Two.

There are a number of other natural materials suited to candlemaking. Any oily object will burn, particularly if fitted with a wick; nuts of most types (particularly the candle-nut (*Aleurites triloba*) or palm-nut of the tropics), birds (the stormy petrel of the Shetland Islands), fish (the candle-fish of Vancouver Island, the dog-fish of Newfoundland), tree bark (the candle tree, *Jacquina pungens* of Mexico) are a few of the natural sources of light. For those who prefer the primitive ways of life, and who wish to live closer to earth's bounty there is tremendous scope for discovery in providing one's own lighting.

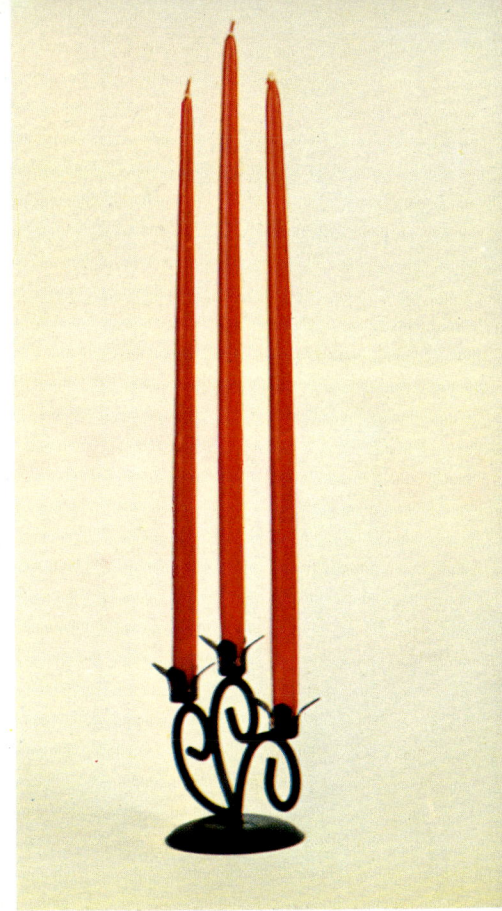

The constituents of contemporary candles

Candlemaking took its first major step forward with the introduction of spermaceti, a by-product of the sperm whale industry which boomed in the late 18th century. In the year 1823 stearin or stearine (actually tristearin is acceptable chemical nomenclature) was first isolated from tallow and then from palm oil (an espressed oil from the fruit of a tropical tree).

In 1825, plaited cotton wicks came into favour, later to be treated with borax solution (sodium borate). Then in 1850 paraffin wax helped the emergence of the snuffless candle, the first really efficient, reasonably priced source of light which was available to the masses. That the last hundred years has shown little change in the formularies for candlemaking introduced in the mid 19th century shows the significance of the progress made in a comparatively few years at the beginning of the 19th century. So great were the advances then made that the materials which were brought into use at that time are those with which we are still concerned and which will now be considered.

Proprietary brands of materials

There are many suppliers, including a number of speciality houses, which sell complete kits of materials for candlemaking. In many ways this is a most convenient beginning, but candles made from kits are more or less standard and do not require a high degree of creativity.

Individual items needed for candlemaking can usually be purchased from the same sources as the kits and in so doing allow more flexibility for the worker to produce original designs. In buying separate items some prior knowledge of usual candle ingredients is necessary, hence, the comments which follow. Most of the ingredients mentioned here briefly are considered in greater depth in the next chapter, therefore minimal information is supplied.

Paraffin wax

Two classes of wax will probably be offered intending buyers, (a) *low-temperature wax* which has some speciality uses, is cheap and imparts draught resistance, and (b) *high-temperature wax* which gives a normally

11

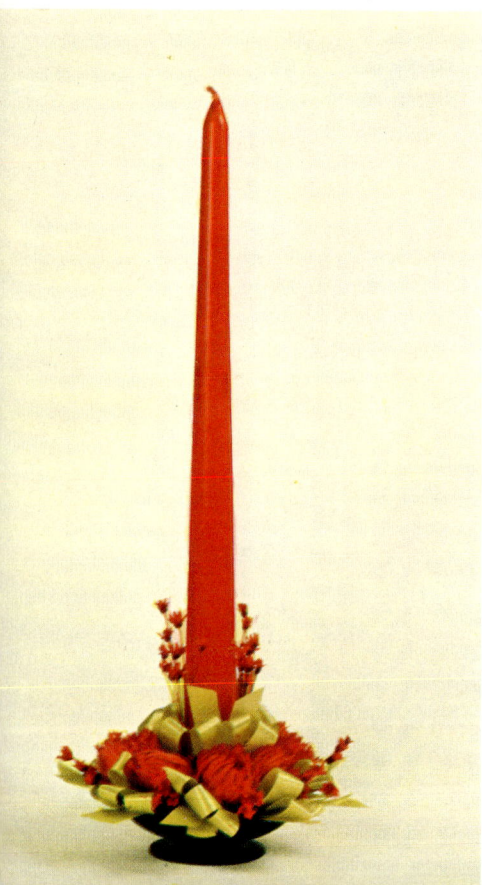

gutterless, long-burning, hard candle. High temperature waxes are the usual choice for straight candlemaking.

Stearic Acid
This is an additive which makes candles harder, longer burning and generally more satisfactory than straight paraffin but the wax gets a less glossy finish and becomes opaque. The loss of the inherent translucency of the paraffin wax through the addition of stearic acid should be carefully considered, as much of the beauty of a lighted candle is caused by the sub-transparent glow below the flame.

Wick
The wick must be matched to the candle, and a decision arrived at as to the proposed diameter of the candle before the supplier is consulted about this item.

Moulds
Commercially made candle moulds are very convenient to use and satisfactory if one is prepared to churn out an endless number of candles of the same size and shape, otherwise they are a waste of money.

Dyes
Much mystery is made out of the need for costly dyes for colouring candles. Certainly, water soluble household dyes are useless for this purpose but there are many other classes of dye stuffs which are eminently suitable. The moral of this is not to be overcharged. Candle dye stuffs are cheaper in bulk than cold water type household dyes and should be comparable in price, if not, someone is making an excess profit out of selling candle dyes.

Perfumes
Perfumery has always been a luxury trade with luxury prices charged. Only a limited range of volatile perfumed oils are satisfactory in candles and these, fortunately, are among the most plentiful and easily extracted of all essences. Do not be overcharged either for these materials. More will be said about this later.

Embellishments
Commercially produced candlemaking kits are offered with a considerable choice of artificial decorations included. Suppliers stock special paints, packaged decorative materials, artificial flowers and like items which are normally charged at a comparatively high price. The best advice regarding these materials is to be wary of spending a large amount of money when alternative low cost substitutes may be available. Later in this book methods of decorating candles are discussed at some length and this section should be consulted before decorative materials are bought.

Candles at low cost

horn lantern

Purchasing candlemaking materials in small quantities or in kitset form is a very expensive way to obtain supplies. This is not completely the fault of the supplier for the cost of breaking down bulk goods into small lots is high so one can expect such items to be costly. In extreme cases, it can be cheaper to buy ready-made candles than to make one's own.

It has been found that those who buy craft materials in the most expensive form, do so because they lack knowledge and have little confidence in their own ability. It is largely in order to overcome these feelings of inadequacy, that this chapter has been written. Like many ancient crafts, candlemaking has been the subject of much secrecy on the part of the masters. The value of many of these closely guarded secrets has proved to be exaggerated but there is a residue of good practical information needed by all candlemakers, whether beginners or experienced.

Basically, a good candle is the result of sound commonsense and an understanding of the principles involved. To some, this chapter may seem overly complicated but the information offered is necessarily technical for if properly understood candlemaking can become an exact operation.

Waxes

The main ingredient in candles is wax and for this reason it is here that the greatest savings can be achieved. Paraffin wax is the cheapest of all waxes but the price asked per pound can vary considerably according to the source of supply, as follows:

1. Sold as candle wax by craft stores and as jelly sealing wax at grocers, paraffin wax is available in 1 lb lots.
2. Sold by oil companies in 60 lb lots (6 blocks, each weighing 10 lb, packed in a carton) the price of the same quality paraffin wax will be approximately one third as much. In other words one can pay the same money for 60 lb of wax from an oil company as for 15 to 20 lb of jelly sealing wax from a grocer.
3. The major oil companies offer at least two grades of paraffin wax with the following specifications being typical of the two classes:

(a) Low temperature range purified paraffin wax
 Melting point 50°–57°C
 Colour near white
 Translucent to sub-transparent and liable to splay, see figure 2/1A.

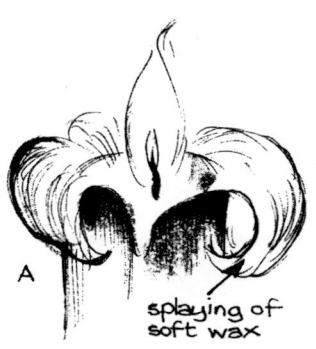

A

splaying of soft wax

Figure 2/1

small flame burning at bottom of deep cavity (cave)

B

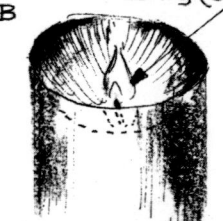

(b) High temperature range micro-fined paraffin wax
 Melting point 60°–72°C
 Colour yellow white
 Translucent, inclined to have a grained appearance. Special clear white waxes of this class are expensive. Liable to cave, see figure 2/1B; costs around twice the basic price asked for low temperature waxes.

4. A candlemaker producing a variety of candles will almost certainly need waxes of more than one temperature range. There are several ways of arriving at a given melting point for candle wax.

There will be few, if any, occasions when a lower temperature wax other than standard paraffin wax is required so it is sufficient to limit the procedures necessary to raise the melting points of paraffin waxes of standard grades. Any melting point between 50°C and 72°C can be obtained by mixing high

snuffer

and low temperature range paraffins in various proportions. This may at first sight appear the best method but in practical candle making this procedure is generally confined to coloured, sub translucent candles owing to the yellowish, grainy appearance of the micro-fined high temperature wax. If a clear white high temperature wax is needed, then it is more economical to obtain this quality by using one of the following methods rather than super-grade paraffins.

(a) Using *ceresin* wax (not previously mentioned) in the proportion of not more than one part ceresin to two parts standard paraffin waxes. Ceresin wax, sometimes ozokerite, is a mixture of complex hydrocarbons fairly closely related to the paraffins.

Melting point	61°–78°C
Colour	near white

Sub-translucent with a somewhat pearly lustre, comparatively inexpensive and readily available.

(b) Using stearic acid in the proportion of not more than one part stearic acid to two parts paraffin or the following recommended recipe:

Standard paraffin wax	6 parts
Stearic acid	3 parts
Bleached bees-wax	1 part

This mixture gives a fine wax for quality candles at a low cost.

Note: To refine and bleach bees-wax: Wax from wild bees and the cappings and discarded combs, which may sometimes become available from a friendly bee keeper can be refined by boiling thoroughly in a quantity of clean water. On cooling, the wax cake is lifted clear and a sharp knife used to trim away the thin layer of bottom wax in which the dirt and debris have settled. For those interested, the honey liquor which remains can be used for making honey mead. Reboiled then skimmed from the surface of the water before solidifying, the wax can be poured into a thin sheet on a clean wooden or glass surface. The yellow colour can easily be much reduced (bleached) by extensive exposure to bright sunlight.

(c) Using up to one part in twenty five parts of synthetic wax with a high melting point. These are sometimes offered by craft shops as 'lustre crystals'. A typical example of these crystals can be bought at a fraction of the price from chemical supply firms as beta-naphthyl benzoate:

Melting point	107°–110°C

Freely soluble in hot paraffin wax

Colour	opaque and chalky white

There are several other possible mixtures and a candlemaker can experiment to obtain individual recipes for different types of candles. As a general guide, (a) and (b) are suited to the majority of decorative candles, and (c) is recommended strongly for taper candles discussed in the next chapter.

Note: Every serious candlemaker must have a thermometer, preferably with a range of from 0°C to 150°C.

Those who may have been appalled at the idea of buying sixty pounds of candle wax can perhaps be reassured by the knowledge that this represents an outlay of less than ten dollars. The next chapter deals with the way in which candles *en masse* can be used, and if the suggestions given there are accepted, then sixty pounds of wax is certainly not too much.

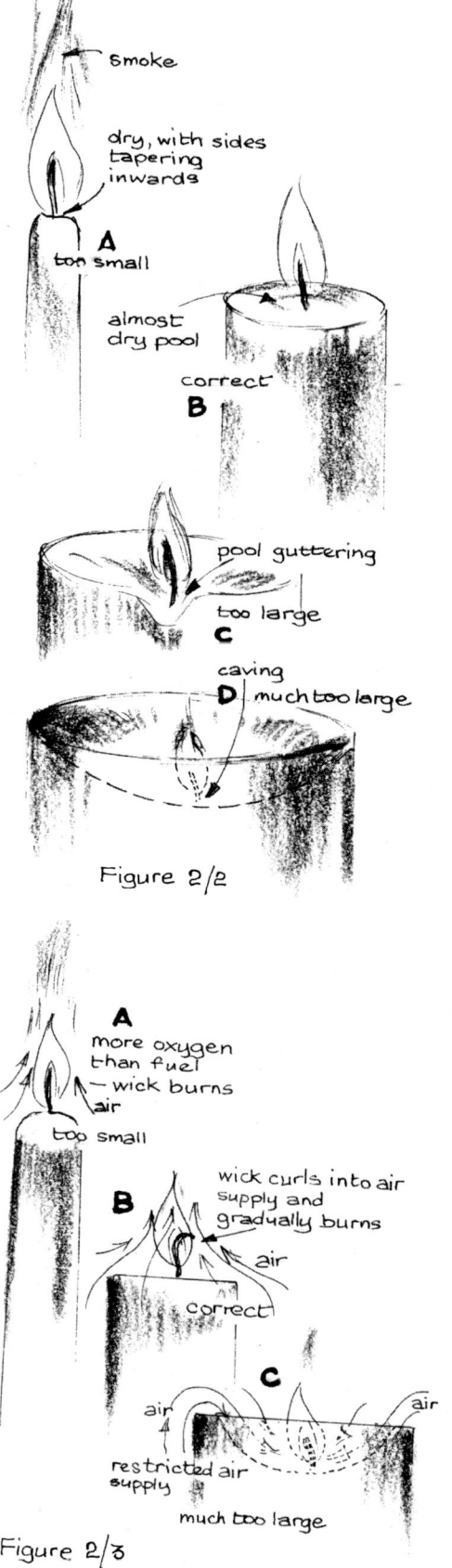

Wicks

No single aspect of candlemaking is more open to argument than the choice of correct wick for a given candle. The difficulties begin with the type of wax used, the size of candle, the purpose for which it is required, and a few other minor factors. Taking everything into account, the most practical approach is by trial and error in order to gain sufficient experience to assess the probable needs of a new candle type.

Trial and error as an approach to wick selection is possible by adding a wick, or a succession of trial wicks, after the candle has been poured. For those types of candle in which the wick must be in place before adding the wax, (tall, slim, taper candles) it may be necessary to make several candles, each with a different sort of wick, in order to satisfy one's self.

Commercially available wicks fall into one of several classes (all based on plaited or braided cotton threads), as follows:

1. *Metal cone wicks* are mentioned first because as well as being the most common type they are so convenient for experimental candlemaking. These have a thin, low melting point wire centre surrounded by the cotton wicking. The wire cone gives rigidity to the wick when pouring wax, in threading an already-poured candle, and when burning. This last characteristic is not always an advantage as will be seen in the discussion on the mechanics of the wick.
2. *Flat plaited or braided wicks* are, as the name implies, of flat construction affording some advantages in air access to the wick. This is another matter dealt with more fully in a later section.
3. *Round and square braided wicks* are the least important of wicks used by the creative candlemaker, and need to be matched closely to the specifications of a particular candle. When standard candle types are in production (for reasons suggested in Chapter 3) then a round or square wick is the first choice.

The mechanics of candle wicks

When a candle flame is first lighted the fuel is either the residual wax in the wick, or when using a new unwaxed wick, the cotton itself. The heat of the flame melts the wax surrounding the wick and provides a pool of liquid fuel (melted wax). By capillary attraction (the same action which causes blotting-paper to soak up ink) the liquid paraffin soaks the wick. The heat boils (evaporates) the liquid, changing this into a gaseous fuel which then burns. The relationship between the size of the flame, the width of the pool of liquid fuel, and the diameter of the candle is critical.

The size of the flame depends on two factors: (a) the amount of liquid fuel available and (b) the quantity of atmosphere oxygen (air) available. The width of the pool is dependent upon the amount of heat (the flame size), the properties of the wax (melting point and a few other factors) in deciding the diameter of the candle.

The effect of these characteristics is shown in figure 2/2 where a same size wick is fitted to each of four differing diameter candles.

Figure 2/2A: There is insufficient wax to feed the wick of the size fitted, hence the wick itself will burn and cause smoking. The top of the candle will tend to taper inwards.

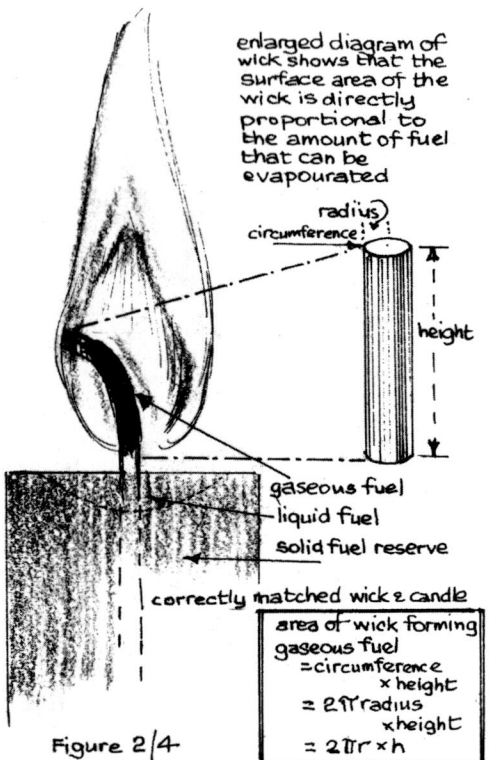

enlarged diagram of wick shows that the surface area of the wick is directly proportional to the amount of fuel that can be evapourated

radius)
circumference!

height

gaseous fuel
liquid fuel
solid fuel reserve

correctly matched wick & candle

area of wick forming gaseous fuel
=circumference
× height
= 2 π radius
× height
= 2 π r × h

Figure 2/4

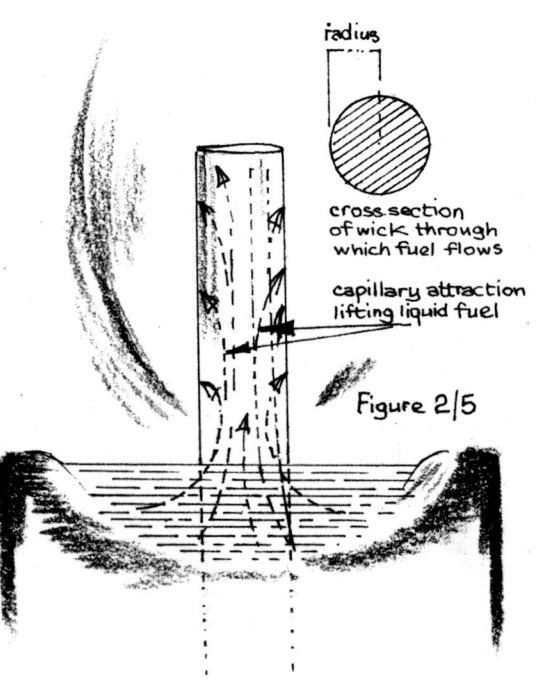

radius

cross section of wick through which fuel flows

capillary attraction lifting liquid fuel

Figure 2/5

Figure 2/2B: When the wick and candle are properly matched the pool of molten wax burns almost dry, showing that the amount of molten wax absorbed by the wick and burnt is almost equal to the amount of wax melted by the heat of the flame.

Figure 2/2C: The over-large candle supplies more wax than the wick can absorb and the molten wax pool overflows and gutters.

Figure 2/2D: The very wide top of the candle is too large to allow the wax to melt to the edges. This causes the pool to 'cave' thus reducing the amount of light emitted. Further, the flame burning low in a cavity has an increasingly restricted air supply which reduces the flame size until the candle stops burning.

Calculating wick size

There are practical limits to the size of an effective wick, as shown in figure 2/3. The air supply is the critical factor examined, and it is candle 2/3C which bears examination. This candle is suffering from a restricted air supply because the wick is not large enough to carry sufficient fuel to absorb all the molten wax available. It would seem that using a larger wick would cure this fault, however, this is not so. Figure 2/4 shows in detail the part played by a wick. Note that the outside area of the wick determines the amount of fuel vaporised.

Consider now figure 2/5, here the amount of liquid fuel is proportional to the cross-sectional area. The cross-sectional area $= \pi$ radius$^2 = \pi r2$. Therefore it is established that the amount of fuel consumed is proportional to $2\pi r \times h$. It is seen that the amount of fuel available $= \pi r^2$.

Restated thus:

Fuel burnt is proportional to $2\pi r$.

Fuel available is proportional to πr^2.

If these geometrical facts are observed in practice such as in figure 2/6, the difference in the formulae for fuel burnt as compared with fuel available can be seen. A correctly burning candle one inch in diameter has a matched wick $\frac{1}{16}$ inch in diameter, (shown as candle A in figure 2/6).

A three inch diameter candle is to be made; what is the correct diameter of the wick? The candle is three times larger and presumably the wick should also be three times greater in diameter, but is this correct? The reasoning can be that Candle A burns correctly so the ratio of fuel available to fuel burnt must be correct. It is to be assumed, as in figure 2/6 that the wicks of the two candles will burn at the same height. While this may be only approximately true, it is close enough in practice to allow the slight difference in the two heights to be ignored.

Hence neglecting the height, the ratio for Candle A will be as compared to Candle B.

Candle A

Fuel burnt is proportional to

$2\pi r = 2 \times 3.14 \times 0.03125 (\frac{1}{32}) \times h = 0.19625\,h$

Fuel available is proportional to

$\pi r^2 = 3.14 \times 0.03125 \times 0.03125 = 0.003066$

The ratio being $\dfrac{0.19625}{0.003066}$ or approximately 64 to 1.

Candle B

Fuel burnt is proportional to

$2\pi r = 2 \times 3.14 \times 0.09375 (\frac{3}{32}) \times 0.38875\,h$

Fuel available is proportional to

$\pi r^2 = 3.14 \times 0.09375 \times 0.09375 = 0.0295$

The ratio being $\dfrac{0.38875}{0.0295}$ or approximately 13 to 1.

If 64 to 1 is the correct ratio (and this is borne out in practice) then the proposed ratio for Candle B, which differs by a factor of nearly 5 (i.e. $5 \times 13 = 65$), must be a long way short of requirements. Those interested may try out other wick sizes. However, such calculations will show that it is not possible to construct a candle of diameter greater than about $1\frac{1}{4}$ inches using a round wick.

Flat wicks

For all block candles, where the diameter is $1\frac{1}{2}$ inches or greater, or alternatively where the distance from the wick to the outside is more than $\frac{3}{4}$ inch, flat wicks are recommended. The plaiting of flat wicks is discussed in the next section. To determine the correct size of flat wick for a given candle it is necessary to remember the ratio of approximately 60 to 1 which can now be restated as a general rule for the calculations of flat wicks when used with candles of one square inch cross sectional area and above. The rule reads:

For every one square inch of cross sectional area of candle, the distance around the wick (the perimeter) is $\frac{1}{4}$ inch linear and the cross sectional area of the wick 0.004 square inch. This rule is illustrated as figure 2/7.

Example

A candle 2 inches in diameter is to be fitted with a flat wick. What is the correct size wick?

$$\text{Cross sectional area} = \pi r^2 \left(r = \frac{\text{diameter}}{2}\right)$$

$$= \frac{3.14 \times 2 \times 2}{2 \quad 2}$$

$$= 3 \text{ square inches approximately.}$$

Required cross sectional area of work $= 3 \times 0.004 = 0.012$ sq. inch.
Required perimeter of wick $\qquad = 3 \times \frac{1}{4} = \frac{3}{4}$ inch.

Practical working

1. Divide the required perimeter value in half $= \frac{3}{4} \times \frac{1}{2} = \frac{3}{8}$ inch
$$= 0.375 \text{ inch}$$

2. Divide this figure into the required area $= \dfrac{0.012}{0.375}$
$$= -0.032 \text{ inch approximately}$$

3. The figure resolved in 2. above is the approximate thickness of the wick. If a plaited wick is to be used, the thread thickness will be one half the wick thickness; in this example 0.016 inch or 16/1000 inch.

With cotton yarns the standard method of 'yarn count' as marked on reels of cotton is approximately:

60 count	0.012 inch—		30 count	0.024 inch—
50 count	0.016 inch—Diameter		20 count	0.028 inch—Diameter
40 count	0.020 inch—		10 count	0.032 inch—

Hence the thread required for this example is 50 count.

4. The width of the wick needs to be 0.375 inch. The number of 50 count threads in this width $= \dfrac{0.375}{0.016} = $ approximately $23\frac{1}{2}$.

However, in this practical example no attempt was made to make allowance for the width of the wick in calculations based on the required perimeter. Therefore, it is possible to subtract $2\frac{1}{2}$ threads from the number calculated and get a more accurate result. Then $23\frac{1}{2} - 2\frac{1}{2} = 21$ threads wide. So the final result is a flat wick of cotton yarn 50 count, 21 threads wide, the specification being 'flat plaited cotton wicking 21/50s.'

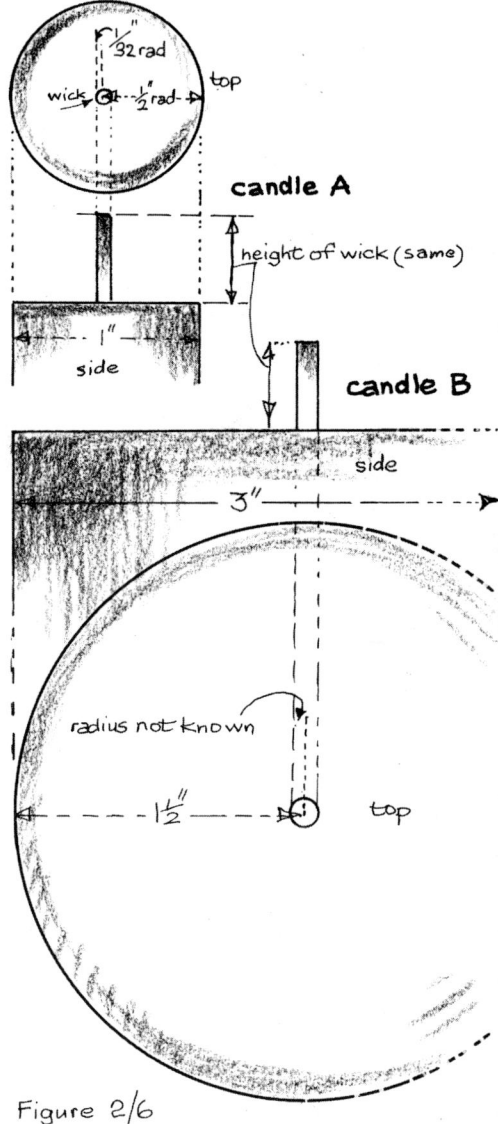

Figure 2/6

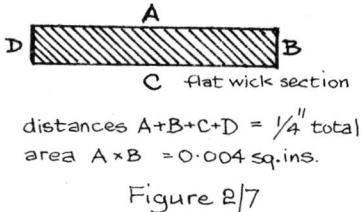

flat wick section

distances A+B+C+D = $\frac{1}{4}''$ total
area A ×B = 0.004 sq. ins.

Figure 2/7

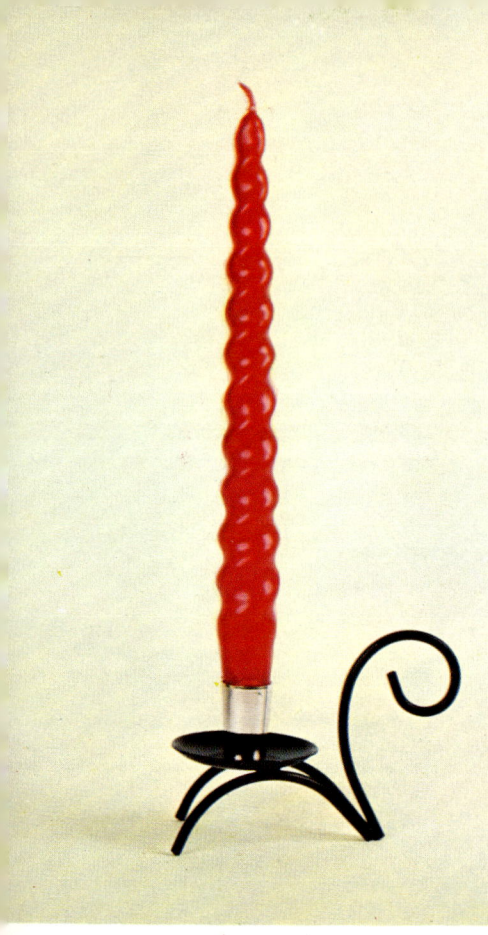

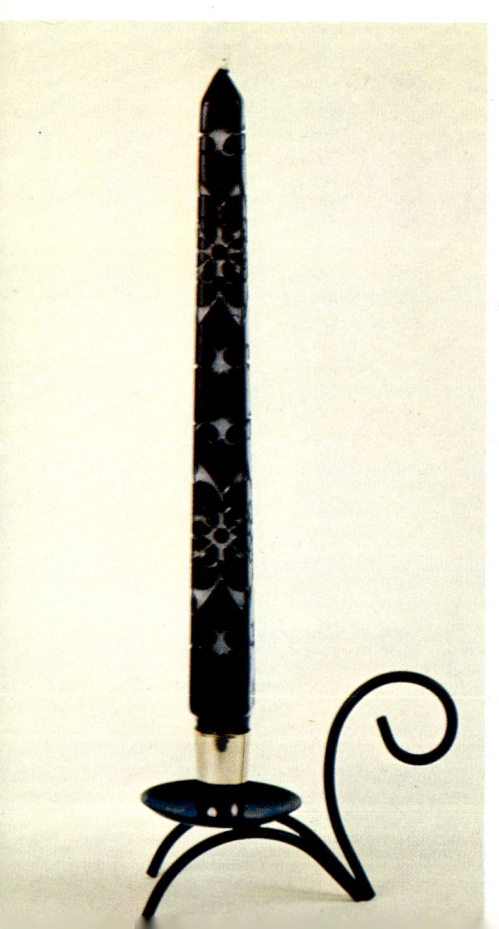

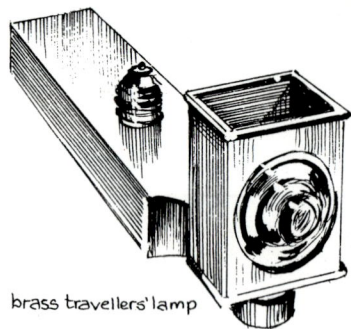

brass travellers' lamp

Choosing wicks (non-technical)

Although the somewhat abstruse calculations just given are essential for understanding completely the technical aspects of wick selection, there is another practical approach which entails a system of trial and error as follows:

1. Decide on a few standard varieties of candles likely to be in heavy demand. For example, it is found that where candlelight has become part of the way of life of a family, 10 inch slim tapered candles and 8 inch, cylindrical candles are used in large numbers.

2. Make several standard candles using a different class of wick for each and select whichever wick gives the best results with reference to figure 2/2. Be sure to notice which candle bears which wick and write this down.

3. With large candles, use a heated ice pick or other similar implement and pierce a wick hole through the length of the candle. It is then simple to thread a metal cored wick or a plain wick, which has been heavily impregnated with wax, through this hole. It does not matter too much if the hole appears disproportionately large for the wick as the molten wax from the pool will fill any cavities. Try several wick combinations in this way and find one perfectly matched to a given candle.

Purchasing wicking

There are many ways in which manufacturers of wick describe their own products. To a beginner this can be most confusing, particularly if buying by mail order. The most practical solution is to order a range of wicking. Fortunately, the material is inexpensive. As a guide, the following details can be offered and the dealer can be asked to supply to these specifications or equivalent. Candles of 1½ inches or less in diameter — size 1/0 woven wick. Candles of over 1½ inches in diameter — size 30 ply flat braided. Candles of over 4 inches in diameter — size 40 ply flat braided. General purpose candles, small — size 34–40 wire core. General purpose candles, large — sizes 44, 24, 18 wire core.

Making wicks

Despite the great importance of choosing the correct wick for successful results the actual process of their making is ridiculously simple, at least for general purpose wicks for candles up to 1½ inches in diameter. Most of the success lies in the chemical treatment of the cotton. Ironically, this fact was discovered and put to general use at a time when candlelight was superseded for most purposes by other types of lamps.

Chemical treatment of cotton wicking — Recipe.

	Parts by volume
Sodium borate (common borax)	2 parts
Sodium chloride (common salt)	1 part
Water	10 parts

Mix well and soak bleached cotton wicking for six to twelve hours, then allow to dry thoroughly. The borax salt mixture is a fire-proofing compound that prevents the cotton of the wick from fast burning, thus ensuring that the wax alone acts as fuel, making the candle more or less self regulating.

Any bleached, undressed cotton yarn or woven material can substitute for regular wicking. Cotton shoe laces, boiled thoroughly in a household bleach solution to remove all traces of 'dressing' and grease make effective wicks. Soft cotton string, candle wicking which is a fabric yarn used for bedcovers, cushions, dressing gowns etc., bears this particular name which must be considered rather inappropriate without further treatment in the context of this book, cotton knitting yarns, dishcloth cotton, and similar cotton products are well suited to making wicks.

For large candles, flat plaited wicks are needed. Figure 2/8 shows the simple plaiting technique employed. Square and round plaits can be used for special wicks and if required 24 to 28 gauge hard drawn copper wire can be utilised as a metal core.

Finally, thoroughly impregnate the wick with wax. Pure bees-wax is recommended and for best results the wicking should be weighted below the surface of the wax which is held for at least twenty minutes in a water bath at around the boiling point of water (100°C). See figure 2/9.

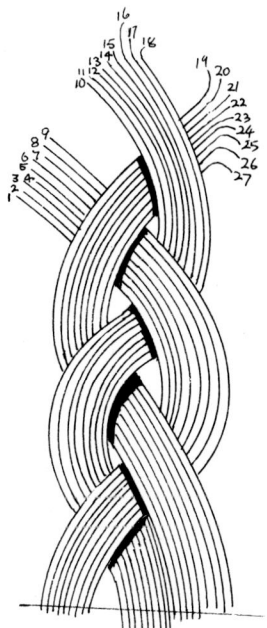

3 × 9 flat braiding
note – any no of threads
can be used giving always
a total of 3 × ......

Figure 2/8

Scents for candles
The usual types of spirit-based or the cheaper, water-based perfumes are not suitable for making scented candles. Special candle perfumes can be bought from dealers, although often they are too high in price.

Delicate perfumes are not really appropriate for candles but there is a whole class of natural oils extracted mainly from plants and dispensed under the name of 'volatile oils' which are most suited to candlemaking. These perfume ingredients are dealt with in full in a companion Creative Leisure book, Glen Pownall's *Perfumery*, together with the making of joss sticks and incense, as well as scented spirit for lamps.

Because that book is readily available it is not necessary to treat the subject of scent in great detail here. However, the following list of essential and suitably robust oils is given and most of them are reasonably inexpensive and have also the characteristics of being insect repellants.

Oil of Balm is from leaves of *Melissa officinalis;* aromatic lemon scent.
Oil of Bay is from leaves of *Pimenta acris;* pleasant sharp odour.
Oil Bergamot is from rind of fresh fruit of *Citrus aurantium;* orange scented.
Oil of Cajuput is from fresh leaves of *Melaleuca leucadendron;* a pleasant camphor-like odour.
Oil of Camphor is distilled from wood of *Cinnamomum camphora;* common camphor smell.
Oil of Cedar Wood is from wood of *Juniperus virginiana;* a pleasant aromatic odour.
Oil of Cinnamon is from leaves of *Cinnamomum cassia;* a spicy perfume.
Oil of Citronella is from fresh grass of the species, *Cymbopogon nardus;* pleasant smelling in low concentrations.
Oil of Dwarf Pine is from the needles of the mountain pine *Pinus pumilo;* a fresh pine smell.
Oil of Eucalyptus is from the leaves of *Eucalyptus globus* and other species; cool, spicy odour.
Oil of Fir is from needles of *Abies alba;* a pleasant, woods-like smell.
Oil of Geranium is from leaves of *Pelargonium odoratissimum;* a somewhat rose-like odour.

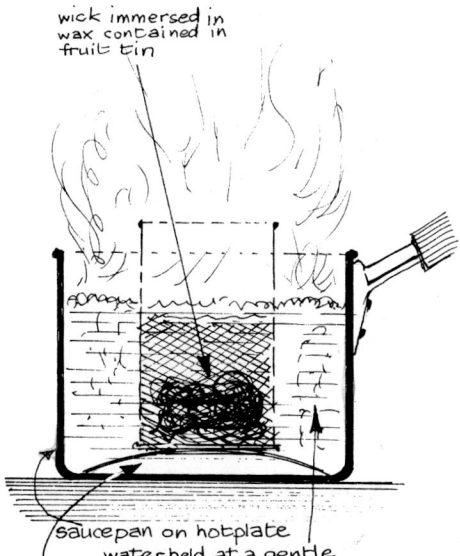

wick immersed in wax contained in fruit tin

saucepan on hotplate
water held at a gentle rolling boil
inverted saucer

Figure 2/9

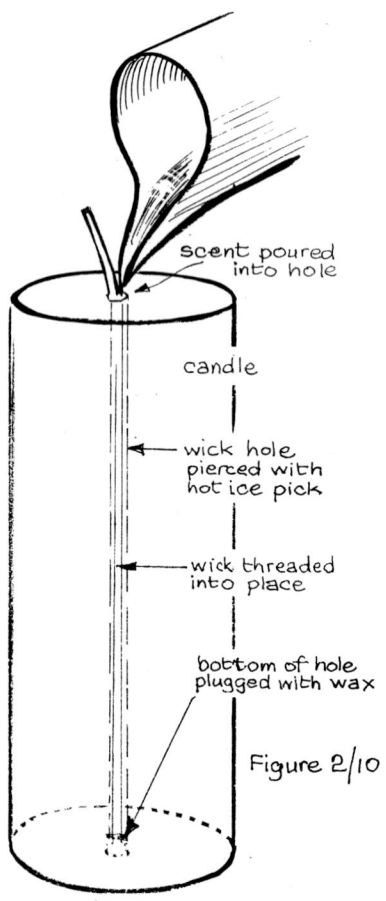

scent poured into hole

candle

wick hole pierced with hot ice pick

wick threaded into place

bottom of hole plugged with wax

Figure 2/10

Figure 2/11

Oil of Lavender is from fresh flowering tops of *Lavandula officinalis;* lavender perfume.

Oil of Lemon is from fresh peel of the fruits of *Citrus limonum;* lemon scent.

Oil of Lemon Grass is from the grass *Cymbopogon citratus;* strong scent of verbena.

Oil of Neroli (sometimes oil of orange flowers) is from fresh orange flowers of the several species of *Citrus;* fresh, highly fragrant.

Oil of Patchouli is from the leaves of *Pogostemon heyneasus;* a persistent, fragrant, oriental scent.

Oil of Pimenta is from the fruit of *Pimenta officinalis;* an agreeable, spicy odour.

Oil of Santal is from the wood of *Santalum album;* a heavy, penetrating oriental fragrance.

Oil of Spearmint is from fresh flowers of *Mentha spicata;* the most pleasant and fresh of the mint smells.

Perfuming candles

All the volatile oils listed in the previous section are fully immiscible (mix freely) with the various waxes used for candlemaking and can therefore be directly added to the molten wax and well mixed. Alternatively, the stronger scents may be directly applied to the wick before waxing or can be poured around a threaded wick as in figure 2/10. Note that the natural, volatile oil perfumes have a strong affinity for waxes, particularly the natural waxes, and this allows a direct method of perfuming candles as follows:

1.   Select a strongly scented foliage or flower. The botanical names of plants which are the source of the commercially extracted oils in the last section can be used. Other alternatives are any members of family Pinus, the white pine, the Himalayan pine, the Pinon pine, the yellow or Ponderosa pine, the Scots pine, the bodgepole pine; either of the redwoods (Sequoia), the Douglas fir (*Pseudotsuga douglassi*) and the cedar of Lebanon (*Cedrus libanotica*) are other trees with highly scented foliage. Family Labiatae (mints) is another group of plants rich in volatile oils contained in the epidermal glands of the leaves and such members of family Compositae as the common sage bush (*Artemisia tridentata*) and other species are also fragrant.

2.   Place a quantity of freshly picked foliage or flower heads as chosen in a muslin bag. Tie carefully and immerse in the hot molten wax and leave for at least thirty minutes, stirring occasionally. The wax will be scented.

3.   To complete the job and make a scented candle look attractive, dip fresh foliage quickly in and out of hot wax and while still plastic stick it to the side of the candle as in figure 2/11.

Colouring candles and decorative wax

Special wax colouring media are available for colouring candles and the wax which is used for decorative motifs. They are expensive to use other than for large scale work, hence possible alternatives are now considered.

1.   One or other brands of dyes as offered for household dyeing of fabrics. All these dyes are powerful and many of them are not suited to wax. However, it is worthwhile to try a few different makes of common dyestuffs, for

their solubility in candle and moulding waxes. For the technically minded it is pointed out that dye-stuffs derived from the benzenoid hydrocarbon grouping are wax soluble, but unfortunately few, if any, proprietary dye products indicate their derivation other than by a trade name.

2. Some brands of wax crayons sold for use by children are very effective for colouring wax when shaved thinly into molten wax. Other, sometimes dearer brands, contain filler material which will spoil the translucency of candles and perhaps clog the wick. Experiment to find the most satisfactory variety of crayon.

3. High quality artists oil colour mixed with a small amount of water-white kerosene can be used for colouring candles.

Note: Any oil, either as volatile oil perfumes or kerosene as a dilutant for oil colours, will tend to mottle a candle. This is not necessarily a disadvantage but must be allowed for when planning special additives to candle wax.

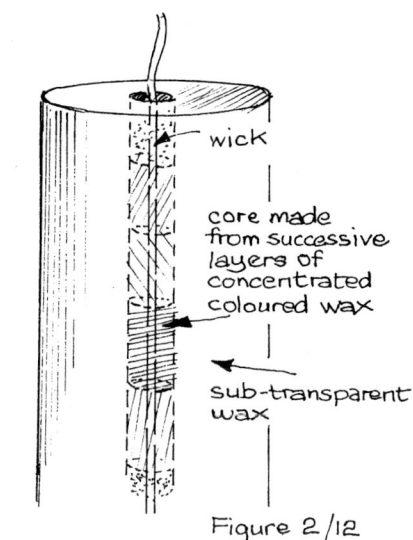

Figure 2/12

Novelty effects

The actual making of candles is described in detail in Chapter 3 where decorations and other non-basic techniques are discussed. However, there are two rather special effects which are probably better mentioned here.

1. Colour drip candles in which several concentrated layers of coloured wax form the multi-layered central core of a sub-transparent candle, refer figure 2/12. The successive layers of colours in the core are laid down in a high melting point wax compared to the temperature of melting of the outer layer. As the coloured wax melts, the molten colour tends to flood the upper surface of the candle and cast coloured light down its main body. If the layers of colour are very thin then the candle will subtly change colour as it burns — a quite admirable effect.

2. By impregnating the wick with different metallic salts, in small sections at a time, the flame itself will change colour as it burns down and so incorporate the different salts with their distinctive colours. The appropriate salts are given in the list which follows. The idea is to make a concentrated solution of each of the salts in a small quantity of water (a concentrated solution is one in which no further salt can dissolve), and then dip small sections of untreated wicking into each solution in turn, as shown in figure 2/13. Unfortunately the wicking cannot be impregnated with the borax-salt mixture as these two sodium salts give off a strong yellow light. This is the reason for using an untreated wick, one which because it lacks fire-proofing impregnants will burn less well than if fully treated. However, the flame will be coloured and will change colour as the candle burns down.

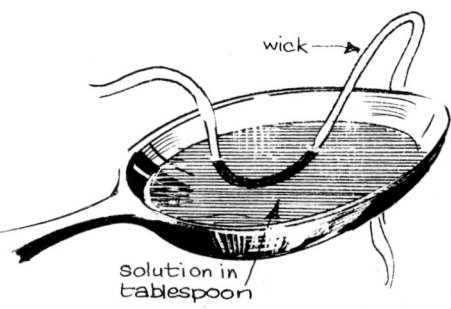

Figure 2/13

Flame colouring ingredients

Any water soluble salt of the metals given below are satisfactory.

Calcium	Brick-red	Thallium	Bright Green
Strontium	Crimson	Arsenic	Pale Blue
Lithium	Deep Crimson	Copper	Blue Green
Barium	Yellow-Green		

This completes the basic knowledge you will need for candlemaking. The next chapter deals with ways of applying this knowledge to greatest advantage.

Massed candles

In terms of lighting technology, this is a sophisticated book, imparting much scientifically accurate information, yet in spite of the knowledge which has been granted to our generation we have yet to match the age of elegance afforded our immediate ancestors. In the great salons of the 18th and early 19th centuries, large numbers of candles were used to illuminate the festivity and revelry of special occasions. These gatherings were the purview of the very rich; crinolined ladies, and bewigged gentlemen bowed and gavotted gracefully in the splendid illuminations of dozens, sometimes hundreds of candles, each group of which had an attendant to trim and maintain the wicks. It is possible, indeed probable, that never again will the world see such scenes. The passing of such opulent living, socially unjust though it may have been, can only be regarded with regret and it is worth considering ways and means whereby at least a part of this past can be recaptured.

Chandeliers of crystal

Chandeliers probably began their history in the Far East where there are still examples of whole rooms walled in facetted crystal and looking-glass mosaic. The light of a single candle within such a room glitters from every wall like a firmament of twinkling stars; twenty candles became a dazzling sight. Crystal chandeliers, those cascades of rainbow-flashing crystal and glass are the Western equivalent of mirror-walled rooms. Chandeliers are made for candles and the diffuse, soft-toned, warm light they emit. As the candlelight flickers, the facets of the crystal 'drops' of the chandelier gather, reflect, and break up the light into countless multi-coloured sparkles and flashes. Because candlelight is warm, which in physical terms means that it lacks the blue-white tones, with reddish hues predominant, a candle-lit chandelier is bright and cheerful.

However, place an incandescent electric lamp in a crystal chandelier and the result is an anachronism. The strong, harsh, concentrated light from the brilliant, white hot tungsten wire of a lamp bulb kills the atmosphere created by a chandelier. An electric lamp giving as much light as forty, seventy five, two hundred and fifty or even greater numbers of candles is not and never will be a substitute for genuine candles.

The last paragraph deals with the prime appeal of candles and if the pun can be excused, highlights a problem that our society has yet to solve, namely, how technology can substitute for good taste, elegance and the inborn urge to live gracefully according to the dictates of human needs.

A single candle, to contemporary society accustomed to high levels of artificial illumination, gives a pathetically poor amount of light for most occasions. As a symbol of intimacy, when supplemented by artfully contrived general lighting, a single candle may create the correct atmosphere, but for the epitome of elegance there is no substitute for a completely candle-lit room.

The production of standard candles

In planning the use of a large number of candles, the economics of candle-making become important. The feeling of freedom to be lavish with candle-light can only come when there is no need for economy, hence first priority must be given to production line methods of producing standard candles.

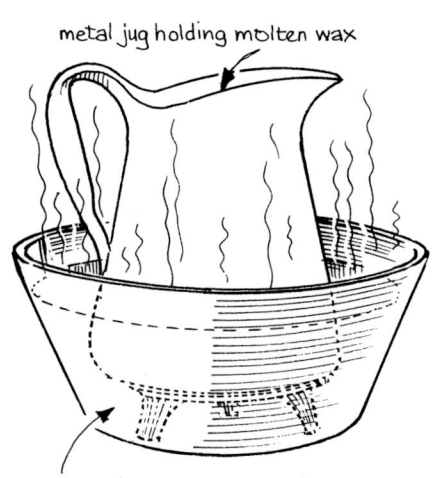

metal jug holding molten wax

water bath containing hot water with jug resting on a trivet

Figure 3/2

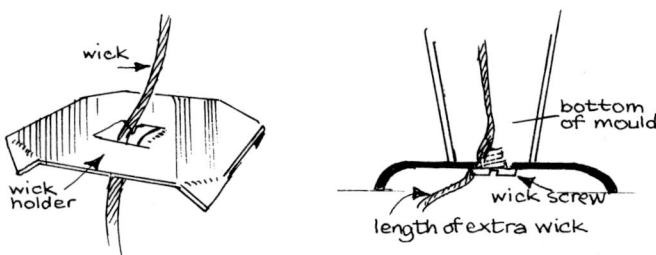

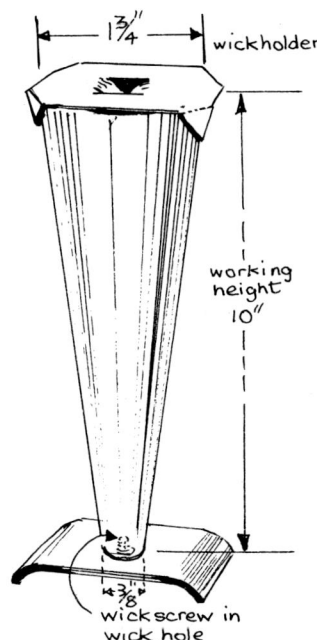

The two specifications given here can be modified to suit individual tastes, but the general technqiue presented should be followed.

Moulds

Several moulds for each type of candle are required. Much of the elegance of standard candles comes from their smooth, clean surface, therefore a heavy, polished, professional class mould should be used. If ordering by mail, obtain the following:

Item one— Three only candle moulds of heavy gauge copper, polished and nickel plated. Inside dimensions being working height 10 inches, diameter at bottom — $\frac{3}{8}$ inch, and height at top — $\frac{3}{4}$ inch supplied complete with wick hole, wick screw and wick holder.

Item two— Three only candle moulds of heavy gauge copper, polished and nickel plated. Inside dimensions being working height 8 inches, diameter at bottom — $1\frac{3}{8}$ inches, and diameter at top — $1\frac{1}{2}$ inches, supplied complete with a wick hole, wick screw and wick holder. Figure 3/1 gives details of an acceptable type of mould.

Melting equipment

For efficient working use a large, easily handled metal container for wax melting. A metal jug or pitcher of at least one gallon capacity is ideal. The heating apparatus requires a pot to give ample clearance around the wax container for use as a water bath, complete with a trivet on which the jug can rest, yet allow ample room for the hot water to circulate under the jug. Figure 3/2 shows the recommended set-up for this purpose.

Cooling equipment

To speed manufacture and create smooth, glossy finishes, a water cooling bath is essential. A large bucket, garbage container or similar utensil can be partly filled with cold water. It is necessary to ensure that the water rises on the outside of the mould to above the inside wax level, otherwise difficulties will be caused by too rapid cooling of one section of wax compared with the remainder.

Candlemaking

There are five basic techniques in general use for the production of candles:
1. Casting wax in moulds. This is the most satisfactory way for amateurs and the method generally advised in this book.
2. Extrusion through a cylindrical die. This is the method by which factory produced candles are made and is generally well beyond the resources of amateur craftsmen. Not recommended.
3. Pouring wax over wicks. A messy, hit and miss way of building up a candle, much used in olden days. Not recommended.
4. Building up by hand. By this technique, highly individual candles can be produced. The method resembles that of a sculptor who builds his work with successive layers of wax or other media. Hand sculptured candles are for the highly artistic, and the technique is discussed later in this chapter.
5. Candle-dipping. A semi-production line method of making utilitarian candles and wax dips. The dipping process itself has much application in

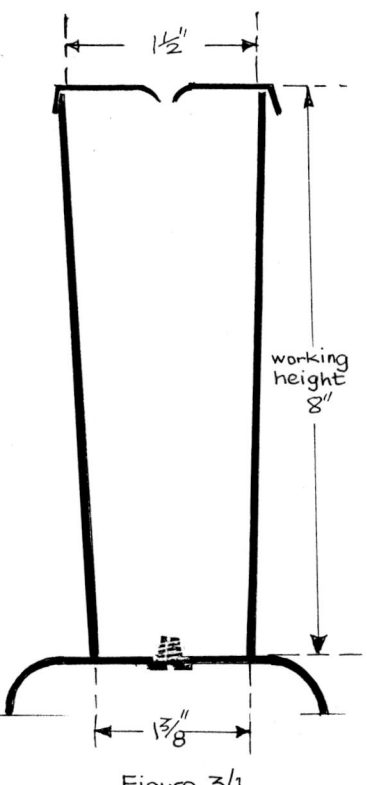

Figure 3/1

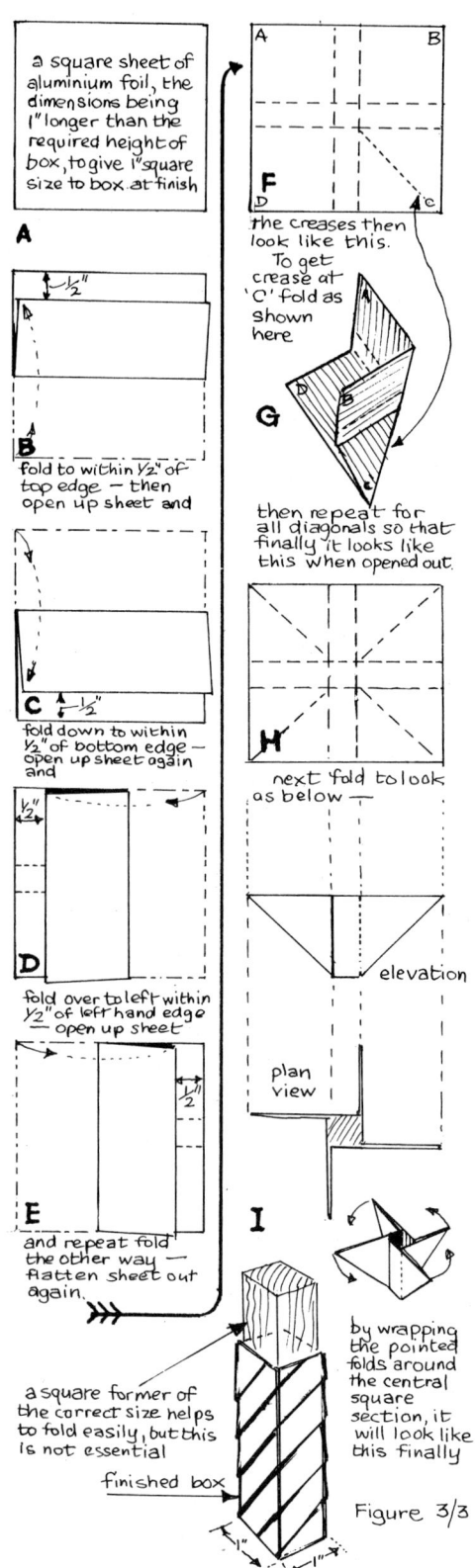

Figure 3/3

decorative work and this will be considered shortly.

Apart from the five traditional processing methods, a sixth may be mentioned; the method of candlerolling, using sheet wax, usually bee-comb foundation sheet. This is specially stamped, thin sheet of bees-wax which apiarists use to encourage bees to build regular shaped honeycombs. Using such a wax sheet highly practical and most attractive, though rather expensive, candles can be made. Candle-rolling is also worth further consideration in this text.

Of the six possible ways of making candles four are now discussed further as each method has its own part to play in the production of candles for special occasions and situations.

Dipped candlemaking

In principle, the making of wax dips and eventually building a sufficient layer of wax to candle thickness is a simple process. In practice, dipping can be time consuming, with control of the finished product difficult. The advantages of dipping however are several and worth enumerating.

1. The equipment required for dipping candles is slight, being only a deep container of molten wax, the necessary wicking and small weight.
2. It is easy to vary the length, and to a certain extent the shape of the candle.
3. It is possible to ensure different coloured layers of wax with little trouble.
4. Possibly the outstanding advantage lies in the production of cores for sculptured, cast and perhaps rolled candles. This point alone makes some knowledge and experience of dipping an advantage to an amateur.

To produce candles by dipping

1. Obtaining a suitable deep metal container for holding molten wax presents some difficulty for the beginner. This is the factor which stops many who would otherwise use this method from doing so. Actually improvised wax containers can be simply made from the heaviest gauge aluminium cooking foil, and these receptacles are very serviceable and cheap. Figure 3/3 shows the method of folding to ensure a wax tight container. The final shaping is somewhat easier when formed around a wooden block of the correct size and shape. These containers are useful for several purposes, and as they can be made to any given size are most economical of wax.
2. To ensure that the molten wax remains liquid the container needs to be held in hot water. Any convenient container is suitable, providing it is sufficiently deep for immersion of the wax pot below the wax level. Wax is lighter than water and the total weight of the foil is not always sufficient to make up this difference. Hence some heavy material, lead shot is ideal, should be put in the bottom of wax pot to ensure it sinks to the correct depth and also to counteract any instability and tendency to topple through the apparent loss of weight when immersed.
3. To form a candle by dipping, a prepared wick a few inches longer than the height of the finished candle should be made ready. A light weight, sufficient to make the wick sink at a reasonable rate through the wax, should be tied to the fuse end of the wick.

24

4. The wax pot is filled with molten wax, care being taken to lower the container into the hot water as the wax is poured. Unless the wax level, from moment to moment, corresponds with the surface of the hot water bath, the wax in the unsupported foil container may cause the sides to bulge. Conversely, if an empty container is forced below the waterline, water pressure may cause the sides to collapse. Therefore if the wax level and the water are the same the two opposite forces will hold the sides in place. As wax is removed during dipping, fresh hot wax must replace this loss.

5. The dipping process begins by lowering the wick to the correct depth and then removing it from the wax. The layer of liquid wax is allowed to harden, and then the process is repeated. If during any of the dippings, subsequent to the first, the partly made candle is allowed to remain in the hot wax, the previously applied coats will melt. Therefore dipping should be fairly speedy and there should be a lengthy pause between dippings to allow the already applied wax to sufficiently harden to resist remelting.

6. A dipped candle can be tapered by controlled dipping to a lesser depth at each dip.

7. If several wax pots, each containing a different colour are used, successive layers of differing hues can be applied. If a clear division of colour between layers is required, then the partly formed candle must be fully hardened between each dip. Even then the baths will be contaminated with other colours.

8. By using controlled depth dips and different colours some rather beautiful effects can be achieved.

9. If a hard, high melting point wax is used for the dipping process, say a wax containing a high proportion of stearin, then this high temperature candle can become the central core of a cast candle made from a wax of a softer and lower melting point. Results can be most appealing, for instance, a coloured core can be seen through the subtransparent, clearer outside coat. The high melting point inner wax will tend to melt a gutter through the outer wax, blending with it and perhaps dripping down the sides in an attractive fashion.

Dipped candles offer ample opportunity for creative work, much more so than the usual run of cast candles and should not be dismissed lightly. The disadvantage is that it is rather boring to just dip one candle at a time. To overcome this difficulty a dipping frame can be made as in figure 3/4.

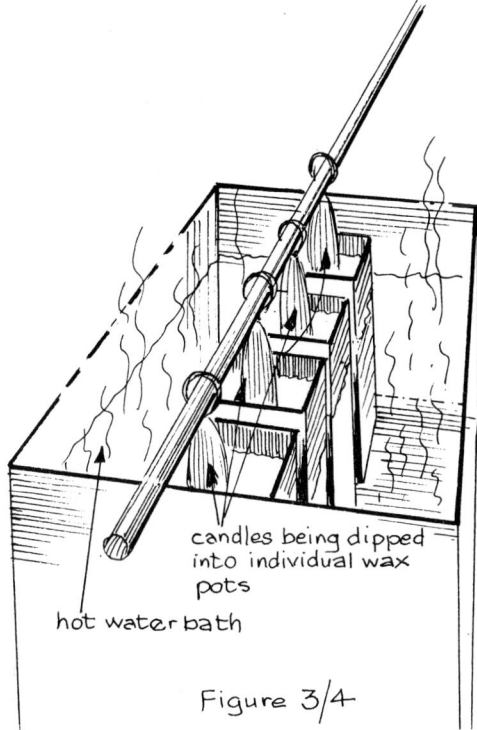

candles being dipped into individual wax pots

hot water bath

Figure 3/4

Sculptured candles

Those with a talent for originality can sculpture their own, unique candles. The process is similar to that used with modelling wax, but in this instance the result is a candle.

A most convenient core model for the foundation is an already formed or partly formed candle, maybe of the dipped variety. Two of the common waxes are particularly suitable for modelling because of their composition; micro-fined wax for its finely worked structure and bees-wax for its special mixture of natural ingredients. Both these waxes will take and hold shape well and will burn readily, with the former being somewhat the cheaper.

Modelling can be a highly diverting way of making attractive candles but there is always the danger that the result will be so satisfying that there is

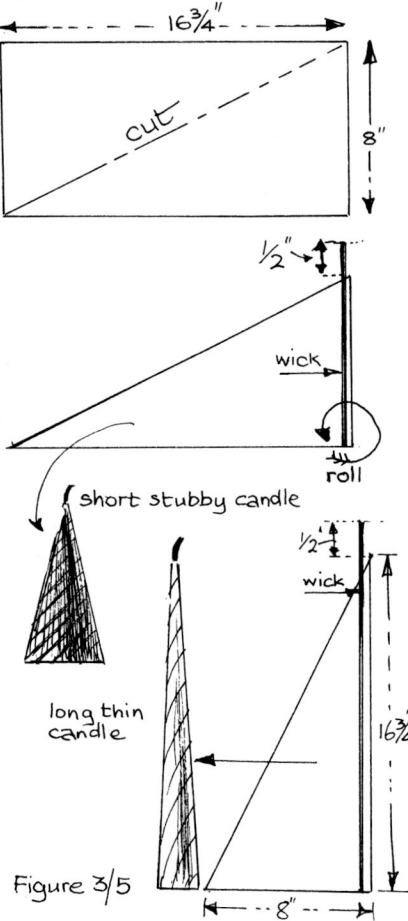

16¾"

8"

cut

½"

wick

roll

short stubby candle

long thin candle

½"

wick

16¾"

Figure 3/5

8"

reluctance to destroy a 'work of art' by burning it up! Possibly the greatest scope for candle modelling lies in the hands of children who can engage in a creative activity without the dangers of hot water and heated wax.

Rolled candles

Prepared bees-wax foundation sheets come in a standard 8 × 16¾ inch size. In the rolling of the sheet an attractive honeycomb pattern is embossed on the surface. To make candles from this material is simplicity itself.

1. Cut a sheet diagonally from corner to corner, figure 3/5.

2. Place a prepared and pre-waxed wick along the edge that represents the height of candle. Allow ½ inch of wick to protrude beyond the limits of the sheet. It will be seen in figure 3/5 that either a thick, stubby candle or a tall, thin one can be made from a half-sheet, as desired.

3. At normal room temperature, roll the candle; it will be found that the natural stickiness of the bees-wax will be sufficient to hold the roll together. The technique of rolling candles is not limited to the production of tapered candles as a full sheet will form a cylinder. Bees-wax is easily moulded by hand and attractive decorations can readily be applied to the surface of the finished candle.

Cast candles (Standard)

Most candles made by amateurs are cast in moulds of one sort or another.

To cast in commercially made moulds:

1. Commercial moulds are not cheap and are readily damaged. The following rules for coring for them must be rigidly enforced.

(a) Be careful not to dent or in any way deform the mould. Even a small dent will lock the hardened candle into the mould and it can only be freed by deforming the finish of the candle in some degree. Therefore, never bump the mould or strike it with a hard object.

(b) Under no circumstances overheat the mould. The surface finish can be damaged by heat, so avoid placing a mould in an oven, say for drying, at any but a low temperature. Never place a mould on a hot-plate or hold in a naked flame.

(c) Do not clean the mould with abrasives or strongly active metal polishes and keep away from all acids and alkalis. When cleaning is necessary, as stubborn wax adheres to the mould, use hot water and mild detergent or some organic solvent on a soft rag. Suitable solvents are petroleum benzine or kerosene, used in the open air away from naked flames. Remember that satisfactory results are only obtainable as long as the mould is treated with the utmost care.

2. Determine the amount of wax required. The simplest way to do this is to pour a quantity of water into the mould so that the level is the required height of the candle and transfer this water into a measure. The wax shrinks between ¼ and ⅓ of its volume between liquid and cooled solid and this must be allowed for when ascertaining the quantity of wax required. One pound weight of paraffin wax will fill, when cold, about 12 fl. ozs. of space. This works out to approximately 1½ standard cupfuls per pound.

3. Set the prepared wick in place in the mould. This wick must be held taut and central in the mould.

4. Raise the wax temperature to about 87° or 88°C, using the thermometer as a check. Wipe off all surplus moisture from the outside of the melting pot. Pour the wax, tilting the mould so the wax runs to the bottom and does not cascade, thereby trapping air bubbles. Some wax will not be used and must be returned to heat so that it remains liquid.

5. Allow the wax to settle in the molten state for about one minute and then place in the cooling bath. Do not let water lap into the mould. (See 'Cooling equipment' section).

6. Keep a watch on the cooling candle, and as the top surface begins to settle into a 'well' shape and a hard film of wax forms, gently break this surface and pour in fresh hot wax. This process must be repeated several times to allow for the contraction of the wax. At no time must the cooling process continue too long with the wax forming a deep film, before 'topping-up' with hot wax, lest the two lots of wax should not blend correctly. The total cooling time will be between one and three hours, depending on the volume of wax used, hence the need for several moulds.

7. At normal room temperature, the candle must cure for at least eight hours before removing from mould. Refrigeration will drastically reduce this time, but a constant check is necessary to ensure the mould is removed from the refrigerator before the contents become too cold.

8. Difficulty should not be experienced in removing the hardened candle from the mould. A smart bump on the base end, with the ball of the hand (after the wick screw has been removed) should allow the candle to slide free. The usual cause of failure to release is an insufficient cooling period. The more sticky waxes, e.g. bees-wax and micro-fined wax may tend to hold, but a quick dip into hot water should free the candle. Several moulds can be in use at one time, and candlemaking then becomes economical in terms of both time and money.

candle holder to wear on lapel of coat 19ᵗʰ century

section shows spring to push candle to top

Candle faults

Faults can develop in candlemaking through lack of care and understanding of basic principles. While such faults will occur in all types of candles they tend to be more serious with the standard type, where purity of form and excellence of finish is all-important. Several of the more likely faults are:

1. *Cavities in the surface.* Probably due to insufficient attention during the cooling period in not breaking the surface film and retopping with wax often enough. If not too deep such cavities can be partly remedied by 'glazing', as considered later.

2. *Mottling.* The candle is not clear throughout the body of the wax. While mottling can be deliberately induced for decorative purposes it is not attractive in isolated areas. It can have several causes; (a) wax not hot enough when poured thus solidifying before all air bubbles have floated free; (b) excess oil in the mixture, added either as a vehicle for dyes or as perfume; and (c) cooled far too slowly, allowing crystal growth within the candle.

3. *Chalk marks on the surface.* Too rapid cooling, over-cooling, or pouring into a cold mould may result in white scaly deposits on the surface. Again, if these are not too bad, glazing will cure the trouble.

4. *Other surface markings.* Numerous small pits, bubbly lines and other

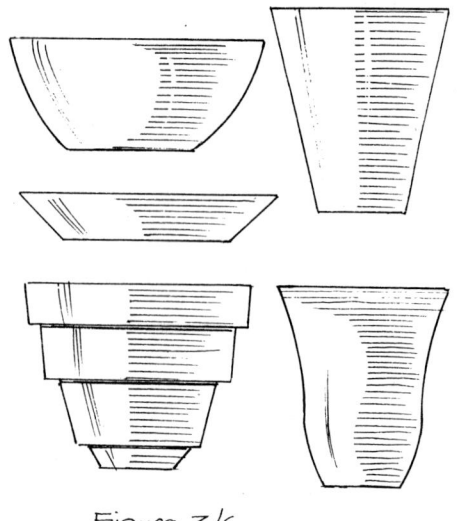

Figure 3/6

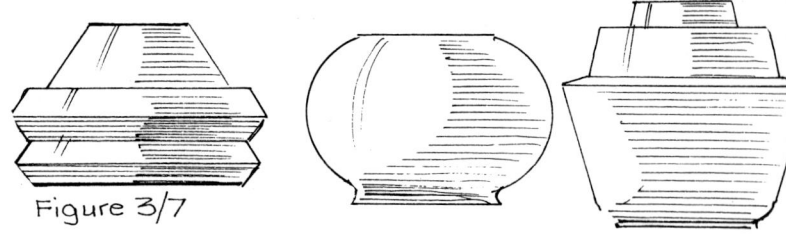

Figure 3/7

blemishes can occur through failure to observe the elementary precautions given earlier. Look at water levels, wax temperature, cooling time and similar matters in order to ensure satisfactory results.

Faults in burning

Spattering, guttering, smoking and failure to remain alight are all caused by the maker failing to fully understand the mechanics of candles as given in Chapter 2, or faultily applying this information in practice. Reference should be made back to the previous chapter, where, at least in theory, all possible causes of poor burning are listed.

Glazing candles

A highly glazed finish with a somewhat liquid appearance can be given to a fully cured candle by dipping it quickly in and out of really hot water. This treatment causes the outer layers to melt and almost immediately to solidify. Alternatively, and rather more effectively, the dipping process can be carried out in hot wax (at 100°C). This latter procedure has several advantages over hot water, as will be discussed under decorative techniques. When skilfully and quickly carried out, a thin, lustrous coating of new wax is deposited by wax dipping which successfully covers small blemishes in the previous surface. Many master candlemakers use this method as a matter of course.

Casting in improvised moulds

Any container capable of holding wax can act as a mould. Generally speaking, such moulds fall into five groups:

1.   Free moulds of such a shape that the candle can be removed. They must be tapered and similar to those shown in figure 3/6.

2.   Moulds which are disposable. They include receptacles of the general shapes shown in figure 3/7. If made of glass the mould is broken to free the candle or if of plastic it is cut.

3.   Multiple moulds which consist of more than one part and which can be taken apart for removal of the candle. See figure 3/8.

4.   Synthetic rubber moulds as shown in figure 3/9.

5.   Free-form moulding in sand or other mediums, see figure 3/10.

   Each of these types of moulds represents a specific example in a varied range of candle shapes. The same general technique of casting candles is used with each group of moulds and where a departure from the basic method is required, this is set out in the following notes.

1.   With improvised moulds the placing of the wick beforehand is usually tricky. It is therefore recommended that the prepared wick be placed in a hole pierced with a slender, sharp, heated tool (e.g. a hot ice pick) after the candle has cured, or that the wick is weighted at the lower end with a metal sinker and lowered through the molten wax before setting. Either of these methods will prove easier and more efficacious than that adopted for commercial moulds.

2.   Plastic moulds should not be subjected to a high temperature, so wax should not be heated above 70°C.

3.   Glass moulds should be preheated by plunging completely into hot

Figure 3/8

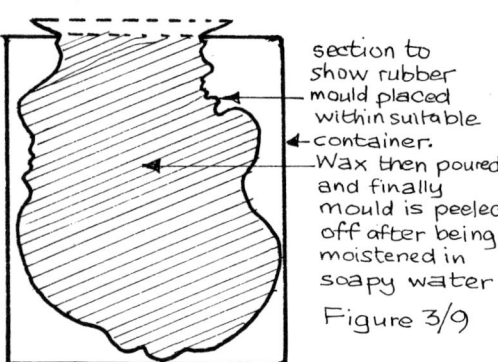

section to show rubber mould placed within suitable container. Wax then poured and finally mould is peeled off after being moistened in soapy water

Figure 3/9

water prior to pouring wax. They should never be placed whilst hot into the cooling bath.

4. To ensure easy release with porous moulds, such as cardboard and wood, swab the insides well with olive, peanut or similar oil.

5. Remember at all times that should wax be heated above the boiling point of water, 100°C, and then poured into a mould containing even a trace of moisture, that the water will almost immediately turn to steam. Steam pressure so developed can drive wax, at a temperature hot enough to cause serious skin burns, out of the mould with some force. Even more dangerous is pouring overheated wax into a container with a narrow inlet. Rapidly cooled wax may plug the escape route for the steam; if so, the mould could explode. Candle making is not dangerous but lack of common-sense, e.g. using an unsuitable thermometer could lead to unnecessary risks being taken.

6. If the design calls for an over-large candle, then it is possible to use more than one wick. See figure 3/11.

free form candle cleared of sand by dipping in boiling water and then cold.

Figure 3/10

Free moulding

A very wide range of suitable moulds is available in most kitchens. A few possibilities are: cooking tins of various shapes; jelly and aspic moulds of both glass and metal; drinking glasses; paper cups; paper cones and any object which will hold hot wax can be used. There should be no difficulty if the general method of pouring as detailed earlier is followed strictly.

Expendable moulds

The essential characteristic of these moulds is that they can be broken or cut to release the solid candle. Into this group fall empty fancy bottles of all shapes in both plastic and glass. These are available today as the packaging of toilet and household items often appearing to be more costly than the contents. Casting candles in such containers is some recompense for paying dearly for fancy packaging.

Glass is a temperamental material. Drop a piece of cherished glassware and it will break, but pound an apparently fragile bottle (especially when supported with solid wax inside) with a heavy hammer and it just won't crack. The simplest and safest way to break the bottle is to pour hot lead over it. The thermal shock and sudden expansion of the glass breaks the bottle with no harm to the candle. Generally the glass will break cleanly without splintering or driving broken glass into the candle. Holding a red-hot poker or other solid steel piece against the glass will usually give the same result.

Plastic can readily be cut with a very sharp craft-knife or scalpel. A thin razor blade may turn over in the hand and is not recommended. The cardboard of containers can be peeled away, layer by layer. This is done most easily if the cardboard is wet.

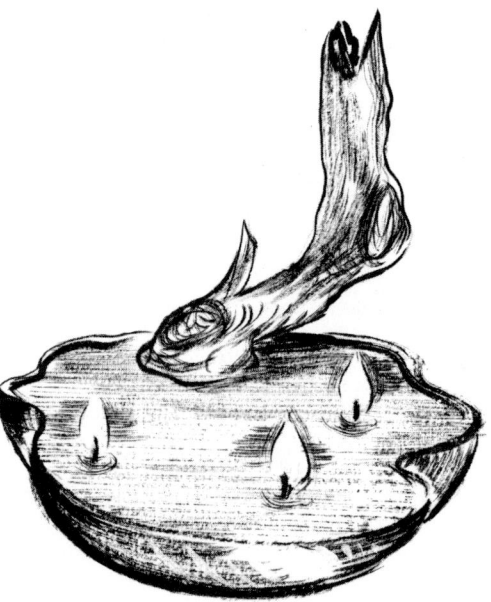

Figure 3/11

Multiple moulds

Traditionally, certain classes of candles have always been cast in two or three piece moulds. Heavy Spanish or Italian candles, deeply embossed with floral motifs, are examples of high quality work of this type. Mediterranean

undercuts
must be
filled in
with wax

wax model

plastic motif

Figure 3/12

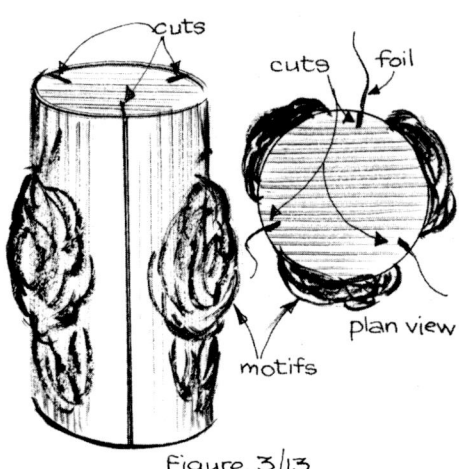

cuts

cuts   foil

plan view

motifs

Figure 3/13

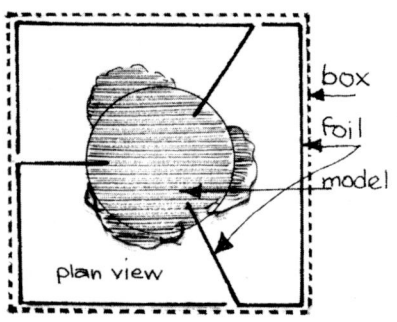

box

foil

model

plan view

Figure 3/14

style candles are still cast in delicate hand-carved wooden moulds, and for a skilled craftsman this method has much to commend it. Plaster of Paris also makes an excellent material for each carving and can be used.

By cheating a little, it is possible as a first attempt to cast a plaster mould using cheap plastic motifs in lieu of hand carving. The method of development is as follows:
1. Use the best materials for the original piece, these are micro-fined wax or bees-wax and are worked by hand (rolling on a flat board when necessary) to the desired shape and size.
2. The required number of plastic flowers, leaves and other motifs are obtained. It is essential that these plastic items are not undercut and if this is so, such parts must be filled in with wax.
3. The plastic motif is warmed slightly and pressed into the surface of the original. The gap between the rear of the motif and the surface of the model is carefully filled in with wax. The necessary filling in is shown as figure 3/12.
4. When the original model is completed, three longitudinal cuts are made in the surface and aluminium cooking foil waxed firmly into these cuts. See figure 3/13.
5. The plastic and aluminium are liberally coated with oil and placed in a suitable container, the model being set centrally and the foil folded around the sides. See   ⋗3/15.
6. Plaster of Paris is mixed and poured into the box and allowed to set for twenty four hours. When dry the mould is removed, whereupon it should fall apart (perhaps with a little assistance) to form a three-part mould.

To use the mould
1. Clean all traces of wax and oil from the face of the mould with petroleum benzine or other solvent and coat with at least three substantial coats of epoxy resin varnish.
2. Lay the mould sections flat on a table and using an eyedropper or similar device fill all the flower parts of the embossing cavity with coloured, high-temperature wax. The colour should be appropriate to the type of flower, and the hue should be bold.
3. Repeat the above but remember the wax should be coloured green where it matches leaves and stems of the motif.
4. Assemble the mould and hold with tight rubber bands.
5. Pour with medium temperature wax, holding the temperature to below or close to the thermometer reading which would cause the wax of the motifs to melt.
6. Complete the pouring operation in the normal way and when cold trim seams with a hot knife.

Candlemaking in multiple moulds is the quickest way of producing a line of highly decorative candles and is often particularly interesting to semi-professional workers.

Flexible moulds of synthetic rubber
Both ready-made flexible moulds and the synthetic rubber composition from which they are formed are available for casting synthetic resin figures, and suffice very well when used with the harder varieties of wax.

To use a prepared flexible mould
1. Support the mould in a suitable receptacle. See figure 3/10.
2. Pour wax in the normal way taking special care to avoid air bubbles.
3. Allow to cool for at least 24 hours.
4. Liberally coat the outside of the mould with a rich, soapy lather and peel the mould back from the candle.

To make custom built flexible moulds, follow the maker's instructions given with the synthetic rubber composition which is available at craft supply stores. Flexible moulds are excellent for casting odd shapes and (as will be mentioned later) for the making of wax motifs for decorating candles.

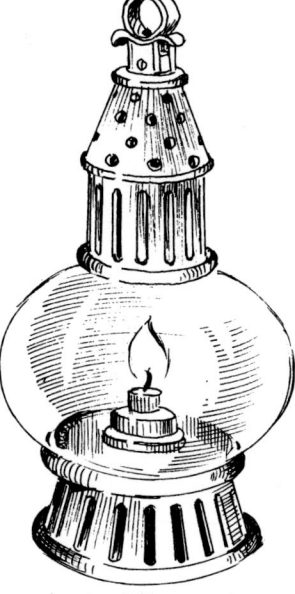

shepherds' lamp c1850

Sand moulding
Highly original free-form candles can readily be cast in sand, salt, vermiculite and other granular material. The method is as follows:
1. Sieve the material through a medium sieve, dampen slightly and place in a suitable container.
2. Make a hole in the sand of the desired shape, using any implement to hand.
3. For a thick sand coating, pour the wax at about 93°C. For a light, easily removed coating, pour wax at a temperature of around 65°C.
4. To remove coating of sand immerse cooled candle in hot water and then cold.

Sand casting has great versatility since unusual effects can be obtained by adding different coloured wax layers. A serviceable sand and wax candle can be made by mixing dry sieved sand to molten wax in the proportion of two parts sand to one of wax. Such candles are heavy and appear chunky but are therefore well in keeping for outdoor use.

Decorating candles
An enormous variety of decorative effects can be applied to candles. The various colour plates, photographs and diagrams in this book show a few of those decorations which imaginative candlemakers can apply. Whereas most of the examples shown clearly demonstrate the particular techniques required, there are a few less obvious methods of producing unique effects.

Chunk candles
These are made basically from cubes of hard, deep coloured wax. The cubes are cemented together with clear, sub-transparent wax; the method is as follows:
1. Add colouring to a high-temperature wax mixture, mix thoroughly and pour into an ice-tray and refrigerate. Allow to harden fully.
2. Remove wax cubes from tray and pack loosly into any mould of suitable size.
3. Heat medium temperature wax to about 110°C, and pour over the wax chunks. The top layer of pieces will slump so several extra wax cubes must be added at the top to compensate for this.

Layered candles
This method of making candles, using horizontal layers of wax, has been

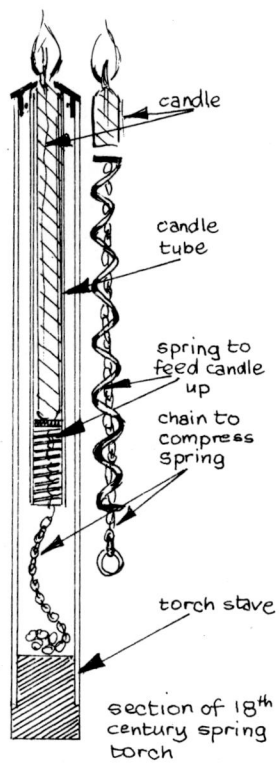

candle

candle
tube

spring to
feed candle
up

chain to
compress
spring

torch stave

section of 18th
century spring
torch

mentioned earlier but other effects that can be obtained have not. The mould can, for instance, be held on an angle, swirling the partly set wax with a stiff wire in order to partially mix the layers; this allows wax of one colour to set against the mould, pouring off the still liquid wax in the centre and replacing with another colour to give vertical layers. Generally speaking, soft pastel shades and sub-transparent waxes should be used for such candles.

Filigree candles
A delicate and unusual candle can be made in the following way:
1. Pour opaque wax into a suitable mould and allow to cool until the wax at the sides has hardened to a thickness of $\frac{1}{4}$ to $\frac{3}{8}$ inch.
2. Remove shell from mould.
3. With the hot blade of a knife cut designs through the wax shell to leave shaped holes into the interior.
4. Replace carved shell in mould and fill with a lightly coloured, sub-transparent wax.

Hand-carved candles
A similar technique to that given for filigree candles requires that a thin shell of wax be first cast, the centre wax poured clear and without removing from the mould, replaced by a contrasting colour. Ordinary carving tools can then be used to carve away portions of the outer shell to expose the contrasting coloured wax. This can result in some striking effects.

Shadow candles
Pursuing the possibilities inherent in the last two techniques allows that wax, paper or plastic decorative pieces be placed inside a wax shell and a new wax core poured to hold their place. Floral motifs are most effective, shining through the wax like very colourful shadows. When putting such pieces in a candle do not forget to allow for the wick being placed later.

Whipped wax
Cooling wax can be beaten to a light, billowy mass with a cake mixer or hand whisk. Allow the wax to cool until a thin film appears on the surface and then beat briskly. The material has wide applications and resembles snow or sea foam according to how it is beaten. Whipped wax, hand-forced into a mould, fitted with a weight on the bottom (lead shots are ideal), and threaded with a wick, makes an admirable floating candle for a centre bowl of an informal table setting.

Candle painting
Small pots of coloured wax kept warm in a water bath can be used effectively for painting candles. Bees-wax is superior to all other waxes for this purpose.

Wax flowers, medallions, etc.
Sheet wax, especially bees-wax, can readily be press-moulded (see Glen Pownall's book *Pottery* in this series for details of press moulding and other pottery techniques readily applicable to candlemaking), or modelled by hand into flowers, leaves, and other shapes which can be stuck on a candle.

34

Drip wax

Whereas a guttering candle is an unappealing sight, a candle with coloured wax dribbling down the side can look very attractive. Controlled pouring of a different coloured wax down the sides of an upright candle is all that is required.

Wax ribbon decoration

An extension of the drip wax technique allows a ribbon of fast-cooling wax to adhere to the side of the candle. Several different coloured ribbons can be applied in any chosen pattern. Before the ribbon has fully hardened, it can be further worked on by hand or tool.

Non-wax decorations

Almost any form of decorative medium can be applied to candlewax. Spirit-based stains as used in woodwork can be employed to give an aged look to the design. Nail polish is an excellent paint for small areas. Glitter, sequins, braiding, paper cut-outs, live plant pieces, artificial flowers, costume jewellery, sea shells, doilies and decoupage, dried plant material, Christmas bells, fir cones, holly leaves and indeed, almost any novel or beautiful article can be applied as a decoration.

Candlemaking is a creative art in which the production of light sources can either be the epitome of elegance, or the most pleasant form of whimsy. Therefore this text continues to offer a wide range of understanding of creative lighting crafts where candlelight is seen to be an essential part of our lives today.

Mood lighting

CHAPTER FOUR   A fault peculiar to our technological civilization is to mistake efficiency for aesthetic values. In the application of lighting, we have become dominated by the cult of the functional whereby something mechanical that works better is in fact of more worth than a less efficient light source. Such concepts are damaging. Appreciation of beauty is not dependent on technological efficiency as expressed in the relationship between energy input and energy output, as this chapter will show.

To see well is important for at least a part of our waking hours. Eyesight is valuable and irreplaceable and only a fool would jeopardise this sense. Therefore correct illumination levels are vital to the well-being of people when this work demands good light. This function of light is well understood and the necessary knowledge is available to ensure that everyone can be free of eye strain. That lighting should also be aesthetically pleasing is another matter which is well worth further consideration.

It would be a foolhardy writer who tried to give a precise definition of aesthetics in terms acceptable to all. However, when considering artificial light there are enough examples of the effect of illumination levels and characteristics to make the term 'mood lighting' meaningful in this book.

Consider the feeling conjured up by the phrase 'moonlight and roses'. Personal reactions will vary in accordance with past experiences etc., but for most there will be at least a slight stirring of romantic feelings. A psychological analysis of its cause could lead into engaging by-ways of the human mind but it is sufficient to state here that the word moonlight has special connotations of a romantic nature. To help evoke pleasant responses by the control of light sources is the prime purpose of this book and will now be pursued further.

The effect of colour
Nobody denies the considerable impact that colour has on our lives. It is therefore proposed to devote this section to considering the mechanics of colour, for undoubtedly colour has a great influence on mood lighting and is in itself a complex and fascinating subject.

Sense of sight
Our window on the outside world is a set of receiving apparatus usually referred to as 'sense of sight'. The eye itself is only part of this receiver (if the analogy of a radio receiver is used); a piece of efficient apparatus, which is really only involved in the comparatively simple transformation of light waves into nerve impulses. See figure 4/1. The remaining complicated responses which tell us with some degree of certainty what goes on around us are carried out in the brain. The way the brain operates, in this respect, is not well understood even by authorities in this field.

Light, as we know it, is our response to the stimulation of our sense of sight, and the way in which we respond is intimately involved with our own internal feelings, level of responsiveness and a whole complex of other personal factors which are not shared with others. The last statement refers to the possibility of a particular message being received by the sense of sight of one person and not evoking the same thoughts and feelings in another person. This difference between people's reaction to given stimulae is

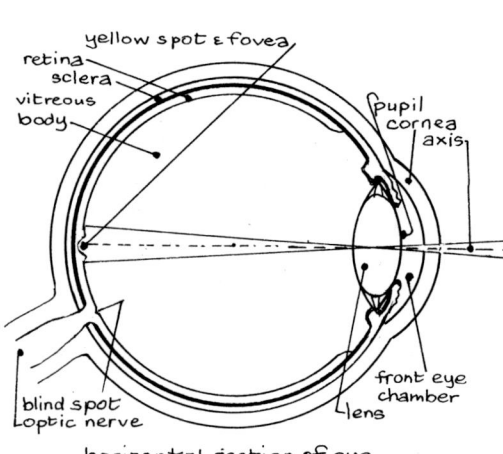

horizontal section of eye

Figure 4/1

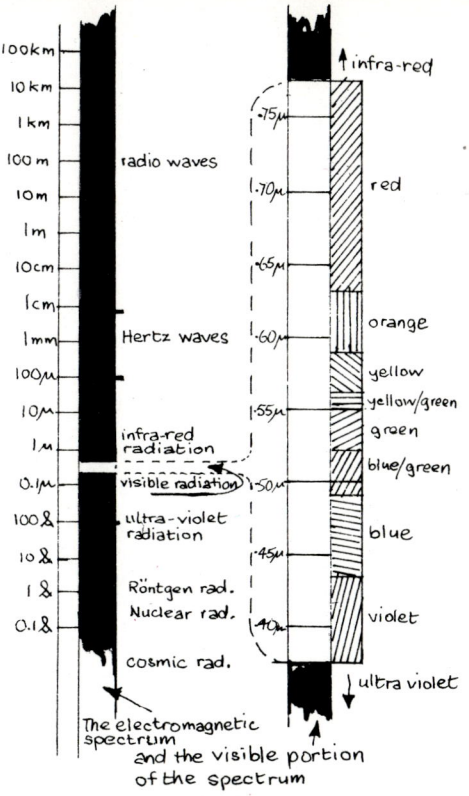

The electromagnetic spectrum and the visible portion of the spectrum

especially true of the psychological effects of colour. This book is not the place to pursue this aspect of colour, but it should be well noted that the remaining text will be tempered by the personal preferences of those involved, and thus it will need to be interpreted according to the needs of those who will make use of it. An example is light of the colour known as magenta; to most people this is a soft, flattering colour; but to a few others it is horrible and repulsive. The difficulty is that both opinions are correct but that it all depends on the minds of the persons concerned. Thus a personal preference is being expressed by an observer.

What is light

This question must be answered in understandable terms, for many difficulties in planning lighting effects arise through lack of knowledge as to the nature of light. Unfortunately, the question is not easy to answer concisely, but can be given thus: visible light, which evokes a mental response through human eyes, is a band of electro-magnetic radiation lying about half-way between the electro-magnetic radio waves used for broadcasting and the highly lethal electro-magnetic radiation given off by an exploding nuclear bomb. See figure 4/2. The visible light which we see in terms of white light is our personal response to the way our eyes are affected by a whole spectrum of electro-magnetic waves, beginning with those of a deep red colour, running through hues of orange, yellow, green, blue, indigo, and violet. Outside our range of vision, are the so called infra-red rays below the red, and ultra-violet beyond the violet.

Though our eyes undoubtedly receive different wavelengths (colours) of light, the effect as interpreted by our brain is that of a white light when all wavelengths are present in the received illumination. In certain circumstances, (such as the conditions in which a rainbow forms) this mixture of all wavelengths of light is resolved into a band of different coloured lights, separated into a number of parallel ribbons of colour. The rainbow is the visible spectrum in which colours must always appear physically in the order of red, orange, yellow, green, blue, indigo and violet. See figure 4/3.

Many substances, both natural and artificial, can absorb or suppress certain of the wavelengths present in white light. If, for instance, all colours excepting the red waves are absorbed (not reflected) by a length of velvet fabric, we say that the velvet is red. In physical terms this is an example of the eye receiving only those wavelengths of light which fall within the red part of the spectrum and the brain interpreting the result as a response to red light. Similar effects are created by all substances known as colour pigments, which give colour to paints, dye-stuffs, the natural colorants of plants, and other materials.

Chromogenes or coloured pigments

By convention, standard white light is taken as being the colour of direct noon sunlight. A perfectly white surface would reflect, totally, all the white light falling on that surface. No such white surface exists, it being found that some light is always absorbed by every surface. The nearest to a perfect reflector is newly-fallen snow in the polar regions, which gives a reflected colour technically classed as a grey of high brilliance. At the other extreme is

spectrum produced by spectral analysis of a beam of "white" light.

Figure 4/3

16th century
lamp

any complete absorber of all light. Such a surface then appears completely black, owing to a total lack of reflected light. A highly efficient absorber of this nature (appearing close to true black) is the pupil of a living eye. All light which enters through the transparent pupil of an eye is trapped inside that organ, and this makes the pupil appear black to an observer. With very few exceptions all other bodies (between the two extremes given above) are chromatic reflectors, being selective absorbers of light. Selective absorption within the visible spectrum, means that a particular chromogene or coloured pigment will absorb some hues more strongly than others, similarly no pigment is a perfect chromatic reflector, some hues other than the desired colour always being reflected; hence pure colour pigments do not exist.

Consider the three primary colour pigments, red, blue and yellow. In theory, mixing these three primary pigments will result in any possible hue, including black. Because of the limitations of selective absorption, no pigment reflects pure colour, disappointing as this may be, therefore a mixture of pigments will never give true black or any true, unmixed colour.

Light colours

When the atomic structure of a substance is stimulated either by strong heating or, less often, by electrical stress, that substance will emit electro-magnetic radiation. If the emission is in the visible spectrum that emitter is said to be a light source. When a body is heated to a high temperature, of about 6,000°K, (K is the abbreviation for the Kelvin temperature scale, which in turn is the celcius (C) temperature plus 273.16°) that body will emit white light. This is not too surprising, as the temperature of the visible and light emitting portion of our sun is 6,000°K.

It is not a coincidence that our eyes are adapted to seeing at maximum efficiency in a visible spectrum produced by a body at a temperature of 6,000°K, as obviously our race evolved under our sun. It therefore follows naturally that our sight is most sensitive to the blue-green wavelengths lying approximately in the centre of the visible spectrum, for through the ages this has been the light available to man, proto-man and our even more remote ancestors. As will be considered later in this chapter, there are profound psychological factors effecting the selective mechanism of the eyes.

From the observation of a rainbow it appears that were it possible to resolve white light into a spread-visible spectrum (as the rainbow in nature) this could be a source of pure coloured light to man. In practice this has proved to be so; man's technological ability has produced pure coloured lights of specific wavelengths, as in the rainbow. The physicist who deals with the fundamental properties of light, specifies his subject not by colour but in wavelengths, giving it a much more precise terminology. For this purpose the unit of measurement is the 'angstrom' with an absolute length of 10^{-12} metre or ten thousand millionths of a metre. Using this physical terminology it is possible to specify the colours of the rainbow as follows:

Indigo/Violet	3,900 to 4,550 angstroms	Yellow	5,770 to 5,970 angstroms
Blue	4,550 to 4,920 angstroms	Orange	5,970 to 6,220 angstroms
Green	4,920 to 5,770 angstroms	Red	6,220 to 7,700 angstroms

According to the generally accepted theory of colour vision, the human eye can receive colour sensations only from red, blue and green lights. Light of other than these three primaries is obtained by adding, in the correct proportion, quantities of two or more of the primary light colours. Hence, yellow light is obtained by adding red and green light although there is in the new colour no wavelength corresponding to the yellow range (5,770 to 5,970 angstroms). The result of mixing light in this way is not a characteristic of light but a consequence of the way in which the eye functions. It is therefore a very human result, which is 'subjective', i.e. caused within the individual. Primary light colours are called additive colours owing to the way new colours are formed by adding existing ones together.

With pigments, taking away (subtracting) colours forms new colours. The addition of a blue pigment to a yellow pigment reduces the total quantity of reflected light. In addition to the loss of yellow colour (as seen by the eye), the blues are now lost as well and what is left is seen as green. Primary pigment colours are therefore said to be subtractive colours.

For an understanding of colour blending and the inducing of moods by colours it is most important to realise that the observer is an essential part of the process. It is stressed that red and green lights do not make yellow light in the true sense of a yellow light falling in the band 5,770 to 5,970 angstroms; neither do yellow and blue pigments make a green pigment. In each case the effect is within the observer; it may be true enough to say that the observer has an illusion of yellow when red and green additive colours are mixed and of green in the case of yellow and blue subtractive colours. See figure 4/4.

a yellow and a blue beam of light superimposed on a white diffuse-reflecting screen produce a white patch

Figure 4/4a

Mixing light and pigment

So far theory only has been discussed, without mention of the possible practical importance of the statements made. There are highly practical matters already dealt with in this text and this is an appropriate place to give examples of the sort of result that can be obtained by colour mixing.

Consider a piece of yellow velvet. It is a clear, brilliant mid-yellow, which in scientific terms reflects most of the light falling on it from the centre of the yellow band, say from 5,820 to 5,920 angstroms. The pigments in the velvet will reflect other colours besides yellow and so the yellow will not be a pure colour. However, when seen in white light the appearance will be overwhelmingly yellow. The material is then illuminated by a pure yellow light. This is quite practical; for instance, a sodium vapour discharge lamp (the fairly common yellow coloured lamps used for road lighting) produces its principal light in a narrow band of wavelengths around 5,890 angstroms. It is not too difficult to screen off all other wavelengths and receive from a sodium vapour lamp a true pure yellow light. Under this pure yellow light, the yellow velvet will appear in true yellow tones for there is no light from other colours to be reflected or diluted.

For the second experiment the velvet is lit with a yellow light formed by the addition of red and green lights. This time there are no true yellow wavelengths of the correct order in the illumination, hence no yellow light can be reflected from the velvet. There are green and red wavelengths of light present and any pigment in the velvet capable of reflecting these two

yellow & blue paint or ink printed over or mixed produce a green "object colour"

Figure 4/4 b

Figure 4/5
colour seen in normal
daylight

the same colours as
above seen by yellow
and blue light only.

colours will do so. The result will be a sickly mixture of colour, far removed from the bright, blazing yellow which appeared under the yellow light.

Exploiting colour
The example just quoted draws attention to the most difficult and important problem that faces anyone who attempts to manipulate colour in order to affect human emotions. To ensure that there is no misunderstanding regarding this problem, a summary of the points covered so far is given.
1.   Colour is subjective in relation to individuals. Our sense of sight in no way approaches the sterile accuracy of a physicist's optical measuring apparatus, so we can be fooled quite readily in our judgment of colours.
2.   The human sense of sight is wonderfully effective in handling and judging subtractive colours (pigments and dye-stuffs) when illuminated by natural white light from the sun. Conversely, artificial light of a limited range of additive colours is new to our experience. Because our sense of sight is not adapted to these new artificial illuminations, we are all highly susceptible to seeing what we want to see and not what is actually there, as proven by scientific measuring devices.
3.   The illusion of colour in an object will change, in our judgment, according to the characteristic of the light which illuminates the object. See figure 4/5.
4.   Not yet mentioned are the very practical consequences of ignorance of the effects of different light sources. This aspect of colour theory is so important that an actual example is now given.

Artificial light and cosmetics
Throughout most of this book, stress has been laid on the glamour of candlelight. Very few readers will disagree but for the ladies it may be interesting to suggest now that more often than not, this same glamorous lighting can make them look like somewhat overpainted harpies. Impossible! Definitely not!
    A few pages back, stress was laid on the fact that our sun produces light at a colour temperature of 6,000°K. Such light is white with a proper distribution of spectrum colours balanced through the red, orange, yellow, green, blue, indigo and violet ranges. Candlelight has a somewhat variable colour temperature which one minute can be as low as 1,000°K and then flare up as high as 2,500°K momentarily. The ordinary electric incandescent light, on the other hand, has a colour temperature of approximately 3,000°K.
    The above figures are significant. As the colour temperature falls, the spectral balance is lost. The lower the temperature, the lesser the amount of light emitted in the green, blue, indigo and violet portions of the spectrum, and the greater the proportion in the yellow, orange, and red bands. This is considered in greater depth later but is worthy of note here.
    With the colour temperature of 3,000°K of an incandescent lamp, there is a considerable change from the true white daylight spectral balance and most people find it difficult to match coloured fabrics under incandescent lighting. With a candle flame there is an even greater change towards a predominance of oranges and reds and an almost total lack of blues and violets in the resulting illumination. Herein lies a catch which should be

known by all beauty conscious women who use make-up. Apply cosmetics under the spectral balance of an incandescent lamp and the frosty blues and greens, while not appearing such true colours as in daylight, still preserve a 'sparkle'. Move into candlelight which is quite deficient in blue colour light and the sparkling blues and greens go drab and dark through lack of the correct light reflections. The delicate peach glow of facial make-up turns a dusky red in comparison to natural skin colours. In extreme cases, carefully applied and well-balanced make-up can appear like charcoal lines and brick dust. This is not an exaggeration as the tiny mirror in a powder compact does not truthfully record what is actually seen. The answer is obvious, if perhaps a little bizarre. When making up for an occasion when candlelight will be used apply the cosmetics under candlelight. That is correct. For an intimate candle-lit dinner party, dispense with the electric light when making-up and use candles instead. A word of comfort — candlelight has sufficient glamour not to require heavy cosmetic applications. It is one of the few light sources where a well-scrubbed 'soap and water' face will have the delicate blush of natural beauty, without the need of make-up. This is not the least of the attractions of candlelight.

The application of coloured lights
In returning to the subject of human reactions to colour changes in lighting, it is appropriate to comment on the lack of information about the relation of genetic factors to observable phenomena. One such field of speculation is human reaction when day turns into night. As the sun sets the quality of light arriving at a given point on earth changes its spectral composition. The red-yellows begin to dominate, and then these too, fade and night arrives.

At twilight, the tempo of life slows. Man, even the city dweller, still feels the age-old pull of nightfall. The day, with its trials and tribulations is coming to a close, and some relaxation and relief from pressing problems should follow. Man has no inbuilt clock. Incarcerated in deep caves in a search for greater knowledge patient observers have lost all sense of time, not knowing whether it is day or night. The feelings normal at twilight are caused by the spectral distribution of the light. Isolated from knowledge of this change and the cycle of human emotion remains unaffected by the passing of day.

The foregoing is not a dissertation on the romance of the twilight hour but a cold factual analysis of man's ancient inbuilt reaction to a change in the composition of light. This knowledge can be applied in practice. As certain clinics for the treatment of disturbed personalities have proved, a light of the correct spectral distribution will soothe and comfort, while a light of different characteristics will arouse frenzied emotions in the self-same patients. Yet most of mankind completely disregards the powerful and profound effects of coloured light upon the personality.

Light crafts
The essence of art is communication, and the artist's job is to convey a message to his audience. The task to be faced in the context of this book is the expression of a mood, an emotional reaction through the manipulation

kerosine lamp

duplex wick

of light. This is a very sophisticated and modern craft, one in which there are yet no masters and so it is the field of the innovator, the creative amateur.

Because there are as yet no real pioneers of the path to success in sculpture and painting in light, there can be no set guide lines, no guaranteed techniques, no sure advice which may be given in order to achieve the desired results. Therefore, the rest of this chapter outlines methods of using light in ways which may produce unusual and artistic effects. The techniques given are experimental, the results uncertain, but it is by putting the ideas into practice that there is much to be learnt and much satisfaction to be gained.

Basic mood lighting

As an experiment it was felt that a relaxing atmosphere could be induced by lighting a lounge with incandescent electric lamps of less than their normal brilliancy. By reducing the input to the incandescent lamps, the designed colour temperature at which they operate would in turn be lowered, proportionately increasing the amount of yellow red wavelengths in this output at the expense of the blue violet. The considerable decrease in total light given by individual lamps, burning at less than normal efficiency would be compensated for by additional lamps. It was hoped that an artificial twilight would then result and induce a mood of relaxation, irrespective of the time of the real twilight. The lounge being fairly long, the first problem arose through imbalance of the illuminations. Hence the necessary illumination levels for acceptable lighting were first considered using internationally recognized standards. These essential preliminaries are now given in full.

Acceptable levels of functional lighting

Of the several technical units of illumination used in the past the lux defined as 1 lumen per metre squared is that mainly accepted today. The implications as to what value of illumination (more correctly, luminous lux per unit of area) a lux represents and how it is derived, is of no great importance here, table 4/1 gives illumination values in lux covering all situations likely to be considered for domestic lighting.

Table 4/1
Illumination levels in homes
SUBSECTION A
Functional lighting on working surfaces (local lighting)
Living-rooms, lounges, etc. (for reading, writing, needlework, etc.)
    from 500 lux minimum to 1,000 lux optimum.
Kitchens (food preparation, service tables etc.)
    from 250 lux minimum to 500 lux optimum.
Bedrooms (mirrors, dressing-tables, bedside lighting etc.)
    from 250 lux minimum to 500 lux optimum.
Other domestic situations (hobby tables, workbenches etc.)
    from 250 lux minimum to 500 lux optimum.
Outdoor (barbecues, games etc.)
    from 100 lux minimum to 250 lux optimum.

18th century timekeeping lamp

Background and accent lighting (general lighting)
Living-rooms and lounges
    50 lux minimum to 100 lux optimum.
Kitchens
    150 lux minimum to 300 lux optimum.
Bedrooms
    50 lux minimum to 100 lux optimum.
Other domestic situations
    25 lux minimum to 100 lux optimum.
Outdoors
    10 lux minimum to 25 lux optimum.

Some analysis of table 4/1 is perhaps in order. There must always be a relationship between the illumination level of local lighting and general lighting, in order to avoid excessive contrast and consequent eye strain. Lighting engineers tend to divide a given room into three lighting zones, as shown in figure 4/6. The preferred ratio of contrast is given in table 4/2.

Table 4/2

Ratio of illumination levels as contrast zones

	minimum	optimum	maximum
Working surface	5	10	20
Surroundings	3	4	6
Background	1	1	1

As an example of the application of table 4/2 consider a living-room with 1,000 lux illumination on the study desk. The acceptable values of background illumination will then be:

(a) Maximum 200 lux.

If the background illumination be greater than the above value there is insufficient contrast between working surface and background and the result is flat and tiring. Hence a minimum contrast of 5 to 1 ratio resolves itself as 200 lux maximum for the background compared to 1,000 lux illumination on the working surface.

(b) Optimum 100 lux.

(c) Minimum 50 lux.

If the background illumination is less than 50 lux then the contrast between the brightly lit working surface and the comparatively dim background is most disturbing.

Mistake number one

The first attempt to impart a restful mood to the room by reducing the illumination level (a natural effect of giving red and yellow tonings to the light by reducing the colour temperature) was completely nullified through not adjusting the level of the several functional lights in the room. It should be pointed out that the illumination given by functional lighting is determined by the levels acceptable in table 4/1, or it is not functional lighting; therefore the first lesson, which is worth emphasising is:

Mood lighting and functional lighting are not compatible and must be dealt with as separate lighting installations even within the same room.

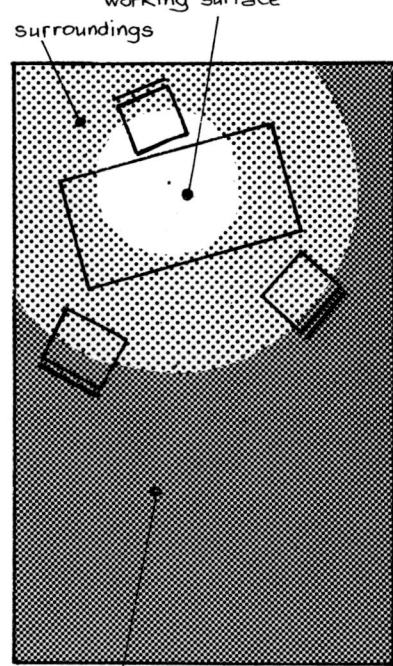

Figure 4/6

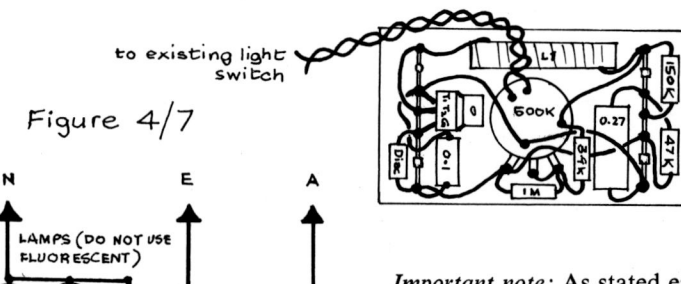

to existing light switch

Figure 4/7

sketch shows possible arrangement of components within switch box.

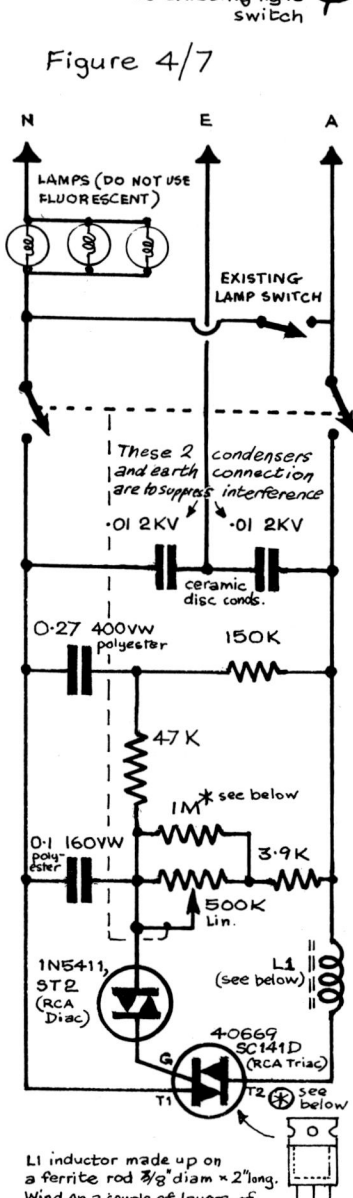

N E A

LAMPS (DO NOT USE FLUORESCENT)

EXISTING LAMP SWITCH

These 2 condensers and earth connection are to suppress interference

·01 2KV   ·01 2KV

ceramic disc conds.

0·27 400VW polyester   150K

47 K

1M * see below

0·1 160VW polyester   3·9K

500K Lin.

1N5411, 5T2 (RCA Diac)

L1 (see below)

40669 SC141D (RCA Triac)

G

T1   T2 ⊛ see below

T1 T2 G

L1 inductor made up on a ferrite rod 3/8" diam × 2" long. Wind on a couple of layers of insulating tape. Close wind 50 turns 22 gauge enamelled wire over tape. Wind tape over the wires tightly in 2 more layers. Buzzing will occur if not wound tightly.

*1M resistor is normal here, but a variation may be necessary in some units to give maximum range of control.

⊛ TRIAC 40669 will control loads up to 300w without an external heat sink. Up to 1000w loads call for efficient heat sink.

Important note: As stated earlier, this book does not deal with functional lighting. As the example just given proves, the rooms of a house fulfil more than one purpose, hence the utilitarian every day needs of living cannot be entirely excluded in favour of purely decorative lighting schemes. If it is desired to update the functional lighting system when installing mood lighting, in order to fully understand the use of such units as the lux, it is recommended that some standard text on lighting be consulted, such as the several excellent publications of the Philips Technical Library, which are readily available. These are a mine of information on matters which are barely touched on here, owing to lack of space.

Basic mood lighting (continued)
The first attempt to provide a simulated twilight having failed, a separate lamp circuit was installed. This was temporary at first, being four wall-type pin-up lamps and two table lamps supplied from a power circuit via a wall plug. Six lamps were required owing to the size of the room and the need for correct light distribution. A proprietary brand of solid state, manually controlled, lamp dimmer was installed in the circuit to control the colour temperature of the lighting. (This type of device is discussed fully later in this section).

Some time was spent in placing the lamps in the correct position with regard to the room and the dimmer regulated to give an apparently reasonable match with twilight. This experiment proved to be another failure and we must digress again in order to understand the reason.

Mistake number two
People become conditioned to certain situations, one of which is to have a room lighted in a manner to which they are accustomed. The 'twilight' room just did not conform to the standards of cheerful lighting expected by visitors and even the designers. The transition from ordinary surroundings into a dimly and statically lighted room was too abrupt. The effect was gloomy and generally akin to that of a not too well lit catacomb! Adjusting the dimmer to the point where a more cheerful situation was produced, cut the device almost completely out of circuit and raised the illumination level to that of a normally lighted domestic room. When compared to the actual setting of the sun, it was apparent that the transition from normal illumination levels to that of a twilight zone should occur gradually and steadily. Manual control of the dimmer proved clumsy and a bother, so the following design was mounted.

Motorised dimmer control
1. A solid state dimmer control capable of handling a lighting load of between 500 and 1,000 watts (equivalent to the use of from 5 to 10 separate 100-watt incandescent lamps) was constructed to the circuit diagram shown as figure 4/7. The circuit diagram includes all the necessary details to enable a competent electronic technician to make the control at a reasonable cost. Craftsmen who wish to make their own will find no difficulty if the components specified (or the equivalent) are used: all joints in the wiring are well made and the materials are solidly mounted, with all live leads shielded.

2. The potentiometer, component 'C' in the circuit diagram which controls the level of input to the lamps, is driven via a short length of plastic tubing from the spindle of a cheap electric clock motor. The plastic tubing is a fairly loose slip-fit on the shafts, so allowing the drive to be released (slip) when the potentiometer arm reaches the end of its travel. See figure 4/8. A return wheel is fitted to the potentiometer shaft to allow the dimmer to be returned to zero when required. This wheel must be reasonably accessible.

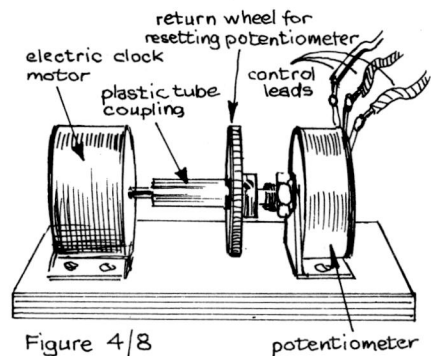

Figure 4/8

Successful twilight lighting

The modified design has proved to be successful and can be recommended as a permanent installation. As a flexible system of background lighting for the room it has many advantages. With the motor not in operation the lighting can be dimmed by hand to any level for television, used as a back-drop to other lighting, as discussed in Chapter 5, and used for mood-inducing effects required on special occasions.

It has been found by experience that a certain sequence of use needs to be followed in order to gain the maximum benefit from the installation; for a dinner party this is:

1. The motor is switched off and the dimmer adjusted to give full output to the lighting. Functional lighting is switched off.

2. Guests arrive and are ushered into a comfortably lighted room with an illumination level to which they are accustomed.

3. The dimmer motor is switched on in the knowledge that the lighting from that moment will be progressively lowered during the next 50 minutes (the potentiometer arm moves through 300° and this takes $\dfrac{300 \times 60 \text{ minutes}}{360}$) to a predetermined level.

4. Cocktails and hors d'oeuvres are served during the next 20 to 25 minutes during which time the colour temperature of the lamps is falling to the stage where it is completely compatible with candlelight.

5. The candles for the table setting are lit.

6. The party joins the table while the background lighting continues to dim, thus enveloping the diners more closely in the candlelight.

The installation, as at present arranged and used, has never failed to relax members of a dinner party and induce a mood of pleasant intimacy. Until brought to their attention, the change was completely unnoticed by guests. Even regular visitors and members of the family who have often been subjected to the influence of the lighting change continue to be pleasantly affected.

Stage setting

The major problem of introducing lighting effects into a domestic scene without introducing a jarring note has thus been solved. It cannot be stressed too strongly that the effectiveness of the new art form of using light as a communication medium is dependent on establishing the mood of the people involved before introducing any unusual effects. The psychology behind this need is fairly obvious and is the reason why this section has been so extensively devoted to what may best be described as 'setting the stage' for the more radical innovations of the next chapter.

Lighting a home interior

CHAPTER FIVE Decorative and mood lighting of a home is an art. It is an art which uses subtle means to set the general atmosphere for a gathering of people and influence others with the personality you wish to project. The last chapter discussed at length how individual preferences in lighting are guided by previous experience and how important it is that the psychological impact of light does not conflict with what people expect to see and feel. This raises the problem of compatibility between human emotions and the quality of light. It is with this problem and the principles upon which a solution may be based, that this chapter is concerned.

of light does not conflict with what people expect to see the feel. This raises the problem of compatibility between human emotions and the quality of light. It is with this problem and the principles upon which a solution may be based, that this chapter is concerned.

Compatibility by candlelight

Certain classes of lighting do not mix. The light from an incandescent electric lamp, for instance, does not mix successfully with daylight. The human eye has adaptability which simply means that should enough time be spent in a light of a certain colour, the eye begins to equate that colour with white. This is illustrated by a blue-tinted television tube which after a short time begins to appear white. So it is with the off-white colour given by an electric incandescent lamp; by itself the colour is equated with white by the human eye. However, illuminate one side of a coloured object with the light of an incandescent lamp and the other with daylight and the result is that the total incompatibility of these two sources is accentuated. In this context incompatibility means that a coloured surface will appear as a different colour in each type of light, and if the lights are mixed there will be an indeterminate blend of hues.

Candlelight does not mix well with either daylight or to a lesser extent with light from an electric incandescent lamp. Dine by candlelight in an incandescent lamp-lit room and the overall result is incongruous; there is too great a difference between the spectral distribution of the respective light sources.

To provide background lighting for a candle-lit table requires some thought. The most satisfactory method is that suggested in Chapter 4 — the reduction of the colour temperature of an incandescent electric lamp to the mean colour temperature of the candles. The solid state dimmer suggested is ideal for this purpose and the matching can be closely controlled as follows:

1. Choose the predominant colour of the furnishings or walls as seen from the table.
2. Place an article of the chosen colour in position as shown in figure 5/1. It is important to so position the test piece that light from each source falls only on one side of it, the opposite side being illuminated by the other light source.
3. Adjust the potentiometer so that the spectral colour distribution from the incandescent lamp contains first more and then less of the upper wavelengths of the emitted spectrum, i.e. the light from the electric lamp is hotter (so containing more green, blue and violet colours) then cooler (so containing less green, blue and violet colours) than the candle flame.

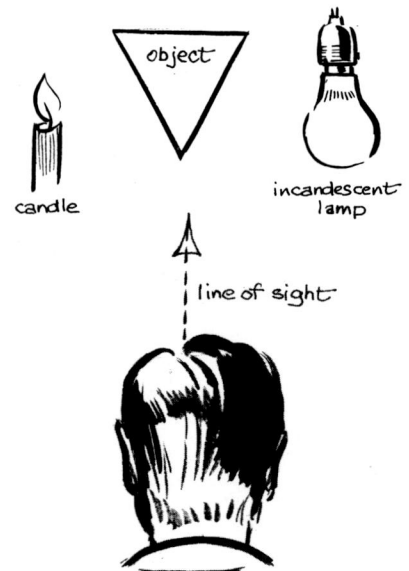

Figure 5/1

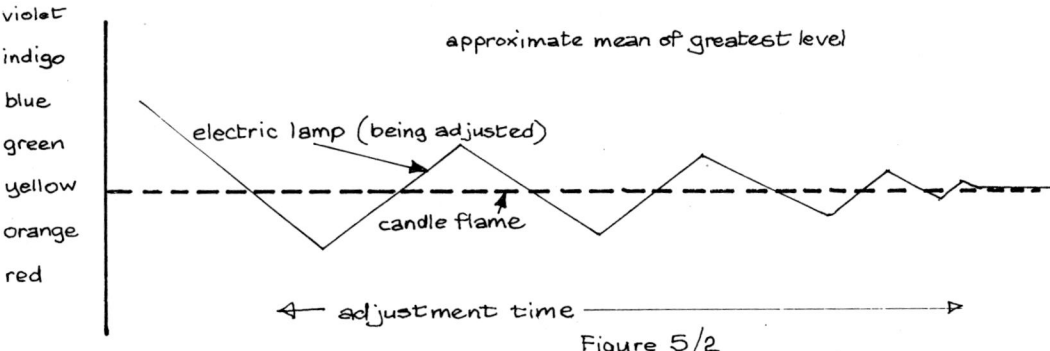

violet
indigo
blue
green
yellow
orange
red

approximate mean of greatest level

electric lamp (being adjusted)

candle flame

←— adjustment time ——————————▷

Figure 5/2

4. The distance above and then below the colour temperature of the candle to which the electric lamp is adjusted is reduced until a close match in the colour of both sides of the test piece is observed. If the adjustment was graphed in accordance with the resulting colour temperature, the graph would appear as in figure 5/2.

Note: Adjusting the colour temperature of the electric lamp successively above and below that of the candle flame tends to minimise the adaptability of the eye and so makes it a poor instrument for this class of matching.

4. The position of the potentiometer arm for satisfactory matching should be marked so that the lamps can always be adjusted to the correct level where the light from the two sources is compatible. Several further points may be added to these suggestions:

(a) The principle of contrast ratio between background and table, as considered in Chapter 4, should be maintained.

(b) If the spectral distribution of the electric lamps is fully compatible with the candlelight the latter need supply only part of the table illumination. This is a convenient way of creating sufficient light at the table without the excessive heat of many candles, which factor could be a disadvantage on a warm night, especially when a number of people are dining in a comparatively small area. It should be noted that per unit of emitted light, an electric lamp is far more efficient (giving off much less heat) than the number of candles required to give the same level of illumination.

(c) If the background lighting is operated at a lower colour temperature so that the deep red hues predominate, the pool of candlelight gives a feeling of close intimacy, apparently cut off from the rest of the room. There are several disadvantages inherent in reducing the colour temperature of the surrounding illumination below a certain level. This level must be ascertained by experiment and it is best not to be too radical in the use of colour before understanding all possible adverse effects.

Colour and colour temperature levels

There is confusion in the way explanatory terms are applied to lighting. A deep red coloration comes from a lower colour temperature source than yellow, green, blue, etc. Physically, the filament of the electric lamp is cooler when emitting red than when giving off other colours. Yet, psychologically, red and reddish hues are said to be 'warm' colours because of the warm feelings they induce. On an evening when the air temperature is getting too high for comfort a psychologically warm light has adverse effects on the occupants of that room. Ideally, a room should have at least two coloured lighting installations, one tending to the warm, reddish hues for winter and cold evenings, the other, for summer being predominantly cooler, based on green-blue tones.

An unbalanced spectral distribution of the illumination will drastically change the colour of coloured surfaces, furnishings, walls, art work, etc. Caution is needed if extreme tones of light are to be used. Red has a peculiar effect (on some people more than others). Approach the red end of the spectrum slowly by similar means to that of the motorised dimmer mentioned in the last chapter and a comforting feeling of pleasant relaxation results; enter directly from normally lighted areas into the same room after

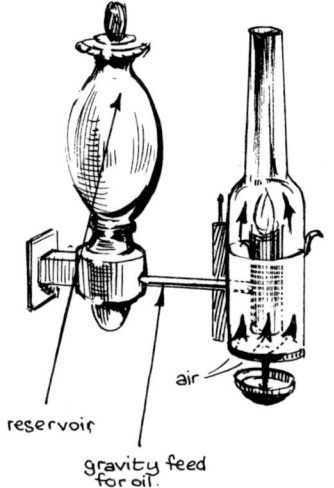

reservoir

air

gravity feed for oil.

Argand lamp

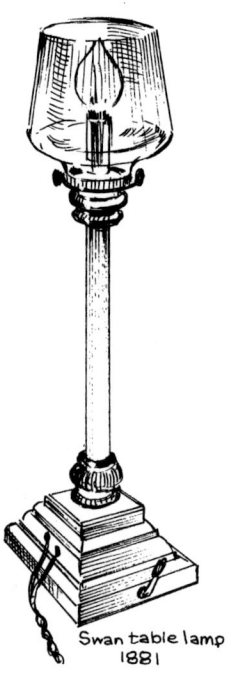
Swan table lamp
1881

the colour temperature of the illuminations has been reduced to its lowest level and the effect can be the direct opposite. Deep red has several unpleasant associations for many people, signifying death, sombreness, the presence of evil spirits and various other archaic and illogical but powerful influences which can be most upsetting. From this it follows that deep red is a colour to avoid.

The lower the temperature at which an incandescent lamp operates the less efficient it becomes. This is not too important in attaining special effects, but the number and rating of the lamps required to provide a certain level of illumination will rise drastically as the colour temperature of the light emitted falls. Against this factor is the near-infinite life of an electric lamp operating at below its rated temperature.

Mismatching by candlelight
To a large extent this section can be classed as the most significant in this book. It could be claimed that any wisdom included in the text is all directed towards the advice now given.

With this in mind, consider the following:
1. The sum of the total knowledge of the effect of coloured light on people is scanty, with far more known of the adverse responses obtained by poor lighting than the pleasures resulting from good lighting.
2. The terms 'poor lighting' and 'good lighting' are impossible of precise definition — a measure of the backward state of the craft of mood lighting.
3. It is known that certain lighting designs can have a profound influence on the mood of persons subjected to them, but the factors giving rise to the mood are imperfectly understood.

This does not amount to a hopeful picture for successful planning of a lighting installation. In practice, the position while difficult is not hopeless, but the point must be made that much more thought has to be given to design and compatibility of light than is now the case.

An increasing number of restaurants and dining places offer dining by candlelight thus copying a popular trend in private homes. The elegant intimacy of such a setting ensures that candlelight will be a pleasant retreat from the stresses of the modern world. Commercial interests have expended large sums of money on the concept of dining by candlelight and individuals are tending to do likewise. It is a fact that many are not successful in creating the intended atmosphere. The reasons are not hard to find; a lack of flexibility and an unwillingness to look at a new problem in a new way. If candlelight is the main attraction of these settings then all other factors are subservient to this and to add candles almost as an afterthought is to reverse one's priorities. The planning is probably carried out in this order:
1. The furnishings are chosen in colours to suit the whim of the decorator. This is the first mistake, for it is almost certain that the tones of the decor are not considered by candlelight.
2. The positioning of the general lighting is planned and then the colours of lamp shades and diffusers are chosen to match the decor. This is the second and major error which should be considered in detail and in view of the following.

(a) Two light sources will have the same spectral composition only when they have the same colour temperature. The spectral relationship between a candle flame and a dimmed electric incandescent lamp has already been considered.

(b) It is almost impossible to match the spectral composition of a light source by using filters, which are semi-transparent materials placed between the light source and the viewer in order to change the colour of the lighting. The usual type of cloth, ribbon, parchment, glass and plastic lampshades when coloured, act as filters.

3.   Finally, candles are placed either in sconces or on the table so that the promise of dining by candlelight is fulfilled.

Under the foregoing circumstances the results of errors become cumulative, including one or more of the following:

(a) The candlelight is not bright enough to provide the minimum ratio of contrast between the table and surroundings, as discussed in Chapter 4. The result is a generally flat and gloomy atmosphere.

(b) The colours of the furnishings, matched to the background lighting lose brilliance in the candlelight, which will certainly be incompatible with the light in which they were chosen. When close to the diner such a mismatch is depressing.

(c) Whatever the level of general illumination, the candlelight, being so near the diners, will compete with the background lighting sufficiently to distort the colours of their clothing and cosmetics. This in itself is unflattering but can be overlooked as the eye is adaptable and compensates for such deficiencies. What is less forgivable is the continual change in appearance with every movement as the amount of light from each source varies on the face and clothing as the position of the body alters. Neither careful grooming nor the utmost attention to complementary colours in attire are proof against the ill-effects of such a poorly planned lighting scheme.

Inherent in the above criticism of this hit-or-miss attitude of most interior decorators is the remedy; begin with candlelight and then plan accordingly. One experience in a room designed on this principle will convince the most sceptical that nowhere else do women look so lovely or men so handsome.

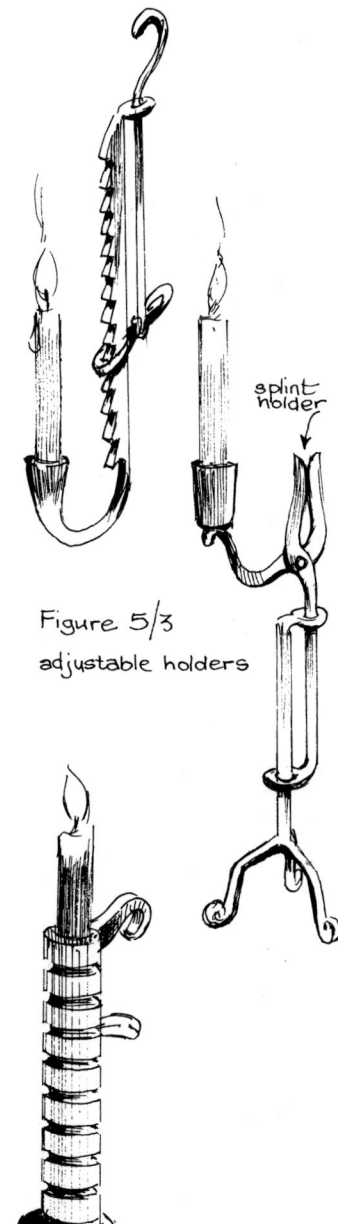

Figure 5/3
adjustable holders

splint
holder

Candlesticks

Not falling strictly within the the subject of lighting is the provision of candlesticks and holders appropriate to the specific setting in which the candles appear. The designing and construction of candle holders is a profitable and rewarding hobby with ample scope for using several talents. It is not proposed to deal with this hobby in this book except to make the following points.

1.   Suitable holders can be made in almost any material and should be sufficiently rigid to hold a shape. Metal and wood have tended to be the most popular materials but pottery, plastic, plaster of Paris, and stone among others, are suitable for novel and attractive holders.

2.   Spiked or pricket candle holders were known before the cupped or socket type and are specially suitable for stubby, contemporary candles.

3.   Adjustable candle holders, as shown in figure 5/3, have a special appeal, particularly outdoors.

Figure 5/5

Roman style lamp

English style
pottery lamp

Kyal lamp,
Cape Cod.
c. 1820

Pennsylvania Dutch
"Cruisie"

twin tube lamp
invented by
Benjamin Franklin

Pennsylvania grease lamp

Figure 5/4

4.   Unless accompanied by a range of period silver and glassware, elaborate candlesticks are out of place today. Candlelighting is in itself sufficiently anachronistic in a contemporary context and to add an over-elaborate holder is likely to detract from general appeal.

5.   Generally speaking, the earlier and simpler types of candle holders as used by the Greco-Roman population in the Mediterranean region and the early Germanic tribes of the North, are more suited to the decor of a modern room.

For the semi-professional candlemaker there is considerable scope for the parallel production of candle holders to be complementary to the style of candles produced. Novelty candles are losing popularity in favour of the more sophisticated classical shapes, and these latter require matching style candlesticks.

Oil lamps for glamour

It has apparently been overlooked by today's designers that oil lamps have a longer history and offer a wider choice of traditional designs than candles. Unquestionably, some of the most beautiful artifacts of all time are the ancient oil lamps found in historic sites all over the world. It is a pity that more use is not made of these.

In order to offer some assistance to those who are interested in the use of oil lamps as items of decoration, the following points are made.

1.   Although old oil lamps were made for the purpose of burning crude vegetable oils or animal fats it is not necessary to use these fuels today. Deodorised kerosene is cheap, readily available, almost smokeless, is fully miscible with perfume oils, and burns brightly without much attention. In every way, with the possible exception of availability, deodorised kerosene is superior to any natural fuel and should be used.

2.   As seen from figure 5/4, the design of lamps in traditional styles varies from the very simple to the really ornate, thus offering a choice of shape to suit every occasion and being within the capacity of any craftsman to construct.

3.   Wicks for these lamps can be made in exactly the same way as that given in Chapter 2 for candles. Actually, the choice of wicking for kerosene lamps is far less exacting than the limits imposed by candles, and there should be no difficulties here.

4.   Lampmaking is of particular interest to at least two specific groups of craftsmen, potters and metalworkers. In view of this it is suggested that the techniques described in two companion books in this Creative Leisure series, *Pottery and Jewellery and Gemcraft*, can be used together with designs shown in figure 5/5 to produce lamps of any degree of complexity.

5.   Specifically and finally, think about the considerable appeal of an oil lamp as a focal point in a properly designed lighting system. The light from its flame is fully compatible with candlelight, and the same criteria apply as have been discussed so far in the chapter.

Contemporary lighting

So far, the text of this book has been directly or indirectly associated with the past. Classical traditions are the refined lessons of history and there is

something healthy about a society which can assimilate classicism into a modern context. From this it can be inferred that present nostalgia is a worthwhile reaction, being not so much a retreat from the stress of today's world, but a need to establish a firm link with our ancestors. Assuming this to be true, then the example of candlelight as a growing part of our social convention is a promising sign of continued stability. However, to neglect the present completely, in favour of the past, can hardly be considered realistic.

Again, the text has, to this point, presented a one-sided view of mood lighting in having covered fully the application of psychologically warm and intimate winter illuminations and neglected fresh and cool summer colours; has offered the demurely elegant and ignored the brightly sophisticated. It is now proposed to make amends, beginning at the level of understanding of mood lighting as so far presented and incorporating lighting techniques well outside the experience of our ancestors from whom we gained our love of candlelight.

Accent in lighting

It is said, by those who presume to know, that every room must have a focal point. If this so-called focal point is to include people, for instance one's partner across the dinner table, then this is true. There are times when a room is not used for dining, or when the pleasant intimacy of being across the dining table is not appropriate. A buffet supper on a sultry summer evening is one such example.

Under these circumstances a room may well require the uplift given by a fixed focal point, an accent in the decor which is noticed, without being obtrusive. Such an innovation can help prevent a slight feeling of disorientation caused by flat, even lighting in strange surroundings.

Accent is a function of lighting. Light is directed onto an object to make that object stand out from its surroundings. Hence any accent in a decorative scheme must be well lighted. This question will be discussed under various headings below.

"Dark" lantern, with door for sudden light —useful for both thieves and police.

Pictures as focal points

Critics say, with some truth, that original Rembrandt paintings, though almost priceless, are seldom displayed to their best advantage. The argument advanced relates back to the purpose for which the paintings were commissioned. For instance, it is said that 'The Night Watch' was intended for a specific hanging position in a candle-lit dining hall. To view this, and many of his other paintings, in a strong light is to exaggerate their tone contrasts and destroy the balance of colour as visualised by the artist.

It is too often overlooked that an artist is concerned to convey his feelings about a specific subject; his personal view of what he perceives. Colour and particularly colour contrasts very often make a major contribution to his communication. This is especially true of some of the latest developments in abstract art. A viewer who sees a work of art in light of a different spectral composition from that under which the artist worked is not seeing that work in the same colours as was intended. It may seem rather ridiculous to suggest that an artist label his work with the specification of illumination under which it should be viewed. However with so many varied artificial light sources available today, the situation is one which should at least be considered.

Apart from the obvious need to avoid too much distortion of the tones of a picture, there is little difficulty in lighting a piece of art correctly. Before siting the lamp it is wise to discover the direction from which the light in the picture appears to be coming. Strangely enough, this precaution is not necessary with pictures in which a feature is made of shadows and strong directional effects; the inherent character of the painting overrides the effects of external lighting. Also it is not necessary to worry too much about the diffuse type of paintings which represents one school of modern art. These paintings do not really need specialised lighting, provided the level of illumination on them is reasonably high.

However, the direction from which the light falls is most important in the case of contemporary, highly-textured works which merge into collages executed in various media, some with quite heavily sculptured bas-relief art. The whole impact of any of these works can be destroyed if the lighting is incorrectly placed. The most deadly effect is created when the light is put central to and at right angles to the surface plane of the picture. Under the conditions (as diagrammed in figure 5/6) the picture is completely flattened and killed. Failing guidance from the artist, it is necessary to experiment with the position of the lighting until the results are fully satisfactory.

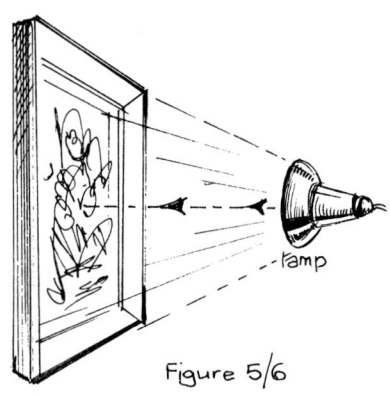

Figure 5/6

Colour compensation

An artifice of which the artist may not approve and which is considered to be unfair by many art critics is to use colour lighting to add brilliance to a work of art. Actually, there are many instances where this way of compensating for lack of colour may be legitimate. A few examples of this will be given here before examining the most successful methods.

1. An old, perhaps treasured painting, may have reached the stage where the varnish has yellowed so much that the colours have become murky. The correct procedure is to have the varnish stripped by an expert, and the painting cleaned. The much less satisfactory alternative, but which in no way endangers the work, is to tinge the lighting with a trace of blue, assuming that blue light compensates in part for unwanted yellow on a surface. The result tends to be more imaginary than real but often proves satisfactory in giving a new freshness to the work.

2. As stated earlier, we are more relaxed under a warm artificial light and so while it is possible to install lighting similar to daylight, we tend to prefer a predominance of red wavelengths in our domestic lighting. Under these conditions a landscape or outdoor scene painted in good faith by an artist working in daylight is seen at less than its true worth. The complete answer is to flood the painting with light from a fluorescent 'daylight' lamp. Automatically, any light spill from a daylight lamp will be incompatible with the normal background lighting and this may cause complications. A less satisfactory but more practical solution is to colour the light from an ordinary incandescent lamp.

3. Until the rise of a well-developed technology in the ceramic industry

certain coloured glazes were difficult to fire true. Yellow was a very difficult glaze to obtain and many beautiful old specimens of yellow pottery appear muddy, especially when heavily lighted. A yellow wash to the colour of the illumination would make considerable difference to their appearance.
4. A subterfuge used with considerable effect in stage lighting is to give depth to a three-dimensional object by lighting each side with different tinted lights. The colder tint will give the appearance of shadows when related to the warmer tint and thus accentuate surface patterns. For heavy, textured surfaces and sculptures this technique can be highly effective in not only giving depth to the exhibit but also direction to a focal point. Two lamps are required.
5. The illusion of movement can be created by several different devices. The mechanics of such illusions will be discussed in Chapter 6 but as the light-modifying characteristics of filters are essential in this application and as filters are discussed in this section the subject is introduced here.

bulls-eye lantern

Colour filters
In Chapter 4 some theoretical consideration was given to the imperfect reflection of colours by pigments. It is necessary to carry this discussion further. When white light passes through a coloured glass plate, a proportion of each group of wavelengths (a spectral group, see figure 5/4) in the spectrum is absorbed by the material of the plate. The smallest amount of absorption will occur around a certain wavelength and the coloured light of the spectral group corresponding to this preferred wavelength will pass through the glass while all other spectral groups will be absorbed to a greater or lesser extent inside the body of the glass. Thus the glass plate is said to be a filter and classified according to the colour of the light which most freely passes through it. Hence a green filter will pass green light, a red filter, red light and so on.

To support this advice, consider the following explanation of the mechanics of a reflecting surface. A near-perfect reflecting surface acts somewhat as a true mirror (there is no such thing in actuality), returning most of the light which contacts that surface. The reflection does not occur at what the eye or sense of touch perceives to be the surface but takes place some distance inside the material. Different wavelengths of light have different penetrative powers which vary according to the characteristics of materials. Hence, reflection from a surface is a matter of short and irregular paths which individual spectral groups traverse when penetrating into, being reflected, and returning through and out of the outer limits of the reflecting material.

Add a pigment to a reflecting surface and each specific pigment is selectively reflective in reflecting a particular spectral group. Hence a pigment which reflects blue light is a blue pigment, and that which reflects red light is a red pigment, and so on. Because of the difference in light paths through a reflective surface, whether pigmented or not, the colour of the reflected light is less pure and less rich, according to the eye, than the same colour perceived as a result of light passing through the set thickness of a coloured glass or other light-transmitting material.

In accordance with the statements made in the last paragraph, it is probable in theory and has been proved in practice that stained-glass windows or other coloured light-transmitting (semi-transparent and transparent) materials give clearer and more brilliant colours than do any reflective materials. Specially manufactured and carefully selected samples of coloured, light-transmitting materials are sold as light-filters. These can be purchased in the knowledge that as products of a reputable firm they will pass a certain section of the wavelengths (spectral group or groups) of the spectrum and suppress all others. This action is technically known as filtering and the item is called a light-filter.

Colouring light
The process of colouring light is the reverse of colouring a fabric or other material. In colouring or tinting light certain spectral groups are removed and the remainder suppressed in order to give a coloured light. In dyeing or staining, certain pigments are added to give a compound mixture of coloured pigments. The essential difference in the two processes is that a distorted colour in light contains too many different colours while a reflecting surface may well be the result of the absence of light of a given colour to be reflected.

Application of colour filters
In the process of applying colour compensation in the possible applications suggested earlier, it is important to realise that its essential object is destroyed if extreme colour filtering is achieved. Take the second example, in which a painting executed outdoors requires some colour compensation in order to balance its colours to something approximating those which were seen by the artist. In effect, it is necessary to illuminate the painting with a light which contains rather less red/yellow than does the raw light emitted by the incandescent source. This is achieved by a very mild, blue filter which displays only a slight preference for blue and thus does not markedly suppress other spectral groups. In this way the blue/greens of the picture are accentuated but the red/yellows not lost. The above rather long explanation is essential if colour compensation is to be correctly understood and applied. Carried out incorrectly with over-compensation, the result is not an improvement but the reverse, a work of art will look far more unnatural than if no filters are used at all. This rather illogical effect is caused by the previously mentioned failing of the human eye to adapt to the colour of the surroundings.

Filters come in a wide range of colours. Some available are yellow, light amber, straw, medium amber, orange, deep orange, primary red, light rose, magenta, ruby, peacock blue, blue green (cyan), steel blue, light blue, dark blue, primary blue, pea green, moss green, light green, dark green, purple, mauve, heavy frost, clear, light frost, medium blue, deep amber, golden amber, pale lavender, pale green, primary green, pale blue, bright blue, pale violet, bright rose, canary, pale yellow, gold tint, pale gold, pale salmon, pale rose, chocolate tint, pale chocolate, and pale grey. It should be noted that each of the filters is compensated for use with incandescent lamps having a colour temperature of 3,000°K.

The final selection of filters for a specific application must always be done on a trial-and-error basis and *in situ*. Until much more is known about the psychological effect of light in different human situations, it is not possible to present any set rules to cover all possible applications, therefore mood lighting must remain an experimental art. Despite this, it is possible to offer a rough guide to the selection of filter colours for the examples given earlier.

1. For compensation for yellowing of the varnish on a painting — steel blue.
2. For imparting daylight colouring to an outdoor painting — steel blue.
3. For giving a clear yellow to pottery glaze — pale yellow.
4. For giving three-dimensional depth — pale gold opposed to steel blue.
5. For movement — a variety of pale tints in tones to match the colour of the object to be exhibited.

Personality through colour
Tinting room lighting to suit the personalities of the occupants is well worth a try. Ways of doing this are simple; use filters to modify the light from the source and very subtly suggest the faintest shade of colour in the background lighting. Accent lights can then throw deeper tones of matching shades which will profoundly influence the mood of the people concerned.

Whereas it is a simple matter to affix colour filters in front of a lamp fitting and so provide coloured light, it is far from simple to choose colouring which is appropriate to a given room and the purpose for which it is used. The atmosphere or 'personality' of a room can be created by colour, and as the layout and furnishings of a home reflect the personalities of those who live in it colour in light can also be used for this purpose.

Warm or cool
The principle of creating a feeling of warmth within a room, by using the warmer spectral groups in general lighting has been fully considered. It may be noted that a series of experiments with average people has shown that, in the opinions of those taking part, the same feeling of warmth is experienced when reddish, warm lighting is used at an actual temperature (as measured) 5°C below that of a cyan-coloured, cool lighting system. In the experiment a red-lighted room at 16°C felt as warm as a cyan-lighted room at 21°C.

The colour known as cyan, is a rather special shade found in nature as a component colour in water, especially turbulent water, as in falls or breaking waves. Physically, as measured by a physicist, cyan is complementary to the primary light-colour of red. The term 'complementary' in this context has a limited technical sense, meaning that a complementary colour when added to the original colour produces an impression of white, as seen by the eye. Hence, from one standpoint, cyan can be considered as the opposite of red in its effect on human perception.

From this it follows that occupants of a room will perceive opposite psychological reactions to cyan light as compared to red light and this fact can be utilised by creating an impression of coolness.

Summer lighting
For most people, living space is restricted, a given room being used for several purposes. The same room will be required to be warm and cosy in

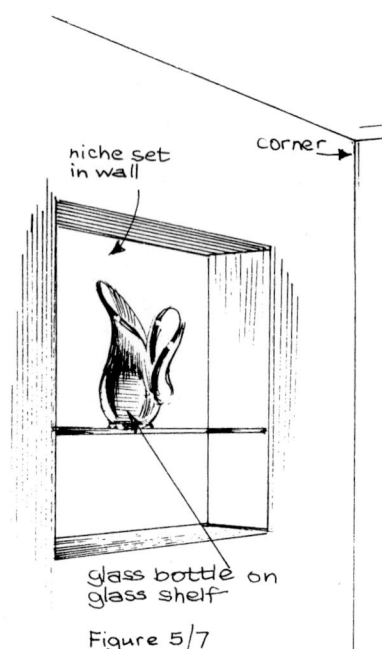

niche set in wall

corner

glass bottle on glass shelf

Figure 5/7

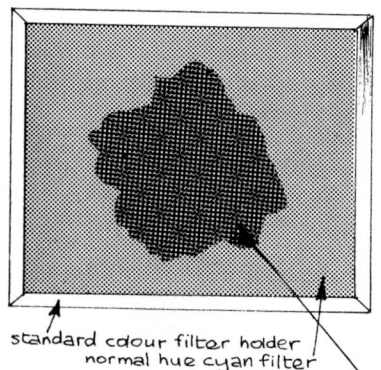

standard colour filter holder
normal hue cyan filter

irregularly cut moss green
filter medium cemented to
cyan filter with clear
nail polish

Figure 5/8

winter, cool and airy in summer; a play and utility room during the day, a gracious reception room in the evening. Therefore, a thoughtful planner will give some attention to the aspect of a room when considering furnishing colours and décor. A room with a sunny aspect will be furnished in cool tones, whereas a shady room is likely to have a warm colour scheme. This is fine, as far as it goes, but it does not go far enough. Consider the following example as to principles involved and the way in which these can be applied in a personal situation.

The problem is how to create an appearance of coolness on a warm and sultry evening. The lighting can be used to promote two illusions which must not cause an icy draught but which can effectively induce a feeling that the atmosphere is several degrees cooler.

1. Fit all the background lighting with thin cyan filters. The operative word is 'all' in respect of the background light fittings, for the effect will be completely spoilt if there are streaks of warm and cool illumination. Thin filter medium is specified, for the toning from a filter is determined by the length of the path the light traverses when passing through the filter. The cyan hue should not be too obvious.

2. An accent light is essential and this needs to be a rather special kind, as detailed below:

(a) A clear green three-dimensional 'object d'art' is essential as the focal point of this display. A simple shape is needed, the actual shading is not too important but the piece should preferably be sub-transparent, or at least translucent, in order to obtain depth. Venetian bottle glass is near-ideal.

(b) The art piece is set on a clear glass shelf fitted into a niche in the least well-lighted part of the room. This is shown in figure 5/7.

(c) A double filament, triple-rated, internal reflecting lamp, while not essential, greatly increases the flexibility of the installation. This class of incandescent lamp is fitted with two independently controlled filaments of different rating. With the low-powered filament switched on, the light output is at a minimum; with the high powered filament operating it is in the medium range, and with both filaments emitting light the output is the maximum. Thus in the one glass envelope is contained the equivalent of three different lamps, each of a different rating. The internal reflector renders any cumbersome system of external reflectors unnecessary.

(d) A special compound colour filter is made as shown in figure 5/8. The main filter medium is cyan (normal hue) and the central portion is moss green. The filter is carried in a standard filter holder as obtained from theatre suppliers. The lamp is sited in a fixed position in order to shine clearly on to the centre of the focal point. The filter holder is set in front of the lamp at sufficient distance to put the irregular outline of the super-imposed moss green medium out of focus, so its edges appear fuzzy.

The desired effect is to produce a feeling of a cool, deep pool of light into which the senses can sink. This can be most impressive, and with the triple-control lamp this combination can be used in the daytime, provided the shades are partially drawn. The impact can be increased by using an extra lamp to throw moving blue shadows across the display. The constructional details of this device are given in Chapter 6.

Personality lighting

With the exercise of a little imagination one can visualise personality in terms of a colour. There is the ice maiden, light blue; the affectionate type, warm yellow; the regal one, lavender, and so on. Based on the provisions suggested in the last section (summer lighting) it is instructive and fascinating to experiment with colour filters in an appropriate setting to determine what 'face' one wishes to show to the world. The expenditure required for a sizeable amount of filter medium for the experiments is not great in terms of the quite astonishing results achieved.

Before leaving this subject of mood influence, one other colour should be mentioned again — magenta. Magenta was used early in the text as an illustration of the psychological effect of certain colours, and it is a very particular class of colour. Magenta is the complementary colour to the primary light colour of green, bearing the same relationship thereto as cyan to red. As green is in the particular spectral group to which the human eye is most sensitive, it might well be expected that its complementary colour has some rather special characteristics as seen by the eye. This proves to be true but human reactions to this colour are not well understood.

Generally speaking, magenta is considered to be a flattering colour whereby a women looks and feels more feminine and a man comes as close to exuding masculinity as it is possible for his particular personality so to do. This characteristic of magenta light is readily experienced but it is quite impossible to explain the reasons for it. All that can be offered is the advice to give priority to magenta when experimenting with personality colours.

Colour and furnishing

Most women and many men pride themselves on their ability to match colours even when a direct comparison of the coloured materials is not possible. It must be conceded that feminine ability to remember hues is often quite remarkable, resulting in accurate judgment of a colour match, sometimes after a lapse of many days. Unfortunately this talent is of no avail when concerned with colour changes due to a light of different spectral composition. Few people, if any, can predict with any degree of accuracy what will happen to a given coloured surface or fabric when it is seen (by the eye) under a different light.

This phenomenon is mentioned here as a further fruitful field for experimentation with ways of discreetly altering the atmosphere of a room as the function of that room changes. The day-time activity has to be used for social relaxation in the evening and create an atmosphere appropriate to each function in turn. This can be achieved by lighting. However, the subject cannot be left without re-stressing that the success or failure of such a scheme will largely depend on the effect caused by light changes on the appearance of the wall, floor, hangings and furniture within that room.

The uses to which the many characteristics of light and coloured surfaces are put depends on the creative powers of the individual. Good luck to those who are prepared to understand and exploit the virtues of colour in their homes; to these artists an entirely new field is opened up in controlling the personalities of those around them.

Paintings and sculpture in light

CHAPTER SIX Via the medium of light manipulation a small group of artists are demonstrating the ease with which an illusion of movement can be given. This skill, called dynamicism, is a mirror of the stresses and strains of today's world and is a sign of our times. Though it appears the very antithesis of much in the text that has gone before, the subject is included here so that the artist who sees himself in the role of the magician can create technological toys that in their limited ways rival laser beams, modulated light sources and other marvels that tempt but are normally beyond his capacity to reproduce.

Memory and movement
One weakness in setting a focal point in a room lies in the inevitable consequence of people having become conditioned to their surroundings and tending to see only those things which depart from their experience (memory) of that situation. In other words a most striking piece of art becomes just another painting of a well-remembered scene if seen often enough in the same surroundings. Conversely, move a chair or other mundane object away by only a short distance from its accustomed place and it immediately becomes the focus of attention of those who remember its usual place.

The solution to this problem is to change the focal point of a room continually. On a gross level, this can be done by replacing one valuable article with a similar one. Intelligent homemakers do this as a matter of course; a flower arrangement is changed, a picture moved, a choice piece of pottery replaced. This gives the room freshness, a feeling of always being new.

In this section, suggestions are offered as to several ways in which change can be executed by exploiting the illusion of movement so readily wrought by manipulating light. One method of successively illuminating an object with different colours is achieved with the colour wheel.

Construction of a colour wheel
To be fully effective and not distracting, the colour of lighting should alter slowly and almost imperceptibly. A correctly constructed and driven colour wheel can ensure that change does in fact occur slowly. This is the procedure:
1.   Obtain an electric clock motor. One hour is a neat segment of time in which a complete cycle of change can take place and the minute spindle of a cheap clock motor is an ideal motive source for this.
2.   Construct the colour disc in the following way:
(a)   Cut two rings of heavy card to an outer diameter of 12 inches with a rim width of $\frac{1}{2}$ inch. See figure 6/1.
(b)   From the same card cut two discs 1 inch in diameter.
(c)   Select several pieces of filter medium of different colours and cut roughly to the shape and dimensions shown in figure 6/1.
(d)   Lay one disc and one ring, concentrically, on a flat working surface and liberally coat the upper sides of both with a quality latex-based contact cement.
(e)   Coat with contact cement those portions of the filter medium which will be in contact with the ring and disc.
(f)   Press the contact surfaces of the filter medium firmly to the ring and disc. Allow the adjacent edges of the colour medium to overlap as shown.

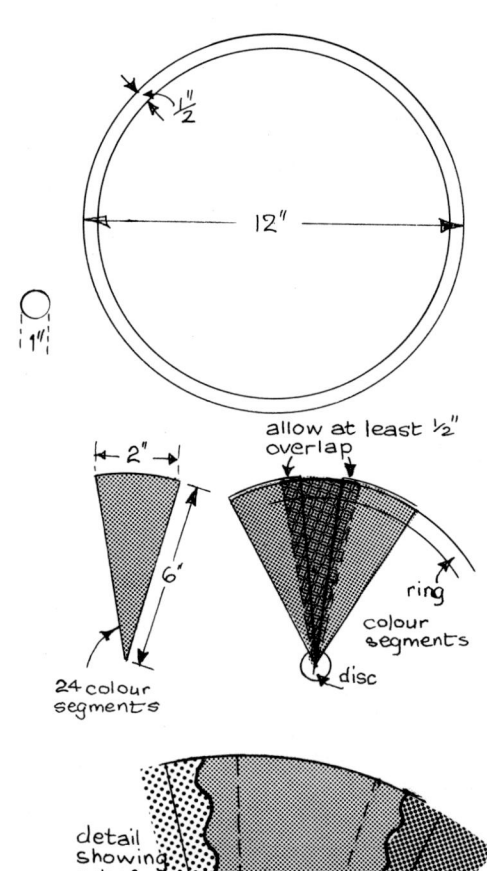

allow at least ½" overlap

ring

colour segments

disc

24 colour segments

detail showing cut of overlaps

Figure 6/1

(g) Repeat the operation, but in the reverse order, to cement the second disc and the second ring over the top of the first pair.

(h) With a very sharp craft knife or scalpel cut through the overlap of the various filter sigments in an edge pattern similar to that shown in figure 6/2.

(i) Centre the disc on the end of a wooden cotton reel and cement in place.

3. The motor and the wheel are mounted as shown in figure 6/2. The colour wheel drive shaft is $\frac{1}{4}$ inch in diameter, hardwood dowelling running in $\frac{5}{16}$ inch holes drilled in pinewood blocks. At the exceedingly low rate of rotation, one revolution an hour, precision engineering is not required.

The colour wheel and lamp are mounted as a unit. The dimensions of the colour disc may need to be increased in order to accommodate a larger type of lamp than that for which this colour wheel was designed. For the sake of appearance, the assembly should be mounted in a case, or alternatively set back at an angle into the ceiling.

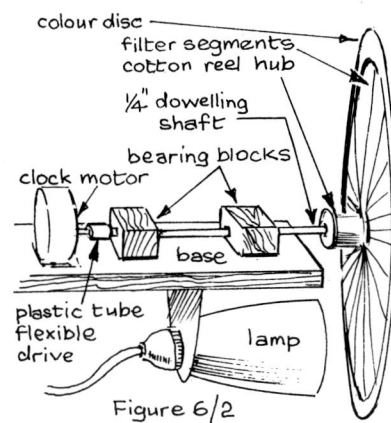

Figure 6/2

Using the colour wheel

A colour wheel is a versatile piece of equipment which can be used for numerous applications, some of which are described later in this chapter. The colour disc has been designed to be readily removed, the normal cotton reel being a neat push-fit on a $\frac{1}{4}$ inch dowel rod. In this way several colour discs can be held and used for different effects.

The shape of the colour segments can be varied, thus altering the pattern of light which is thrown. Worthy of special consideration is the use of two separate colour wheels, so arranged that the light from each falls more or less on the same area. Unusual and striking patterns of colour can be obtained in this way.

Colour programming

Even more flexible and equally versatile are one or other of the systems of programmed switching which can be used to operate multiple lights in a pattern of colour changes. The simplest system of colour programming requires three lamps, one each of red, blue and green colour, preferably internally reflecting. These lamps are mounted so that their light output overlaps on the one display area.

A programme switch is then used to switch these lights on and off in a set sequence to give all the possible colours of the rainbow before the cycle is completed. A sequential programme switch suited to a long period cycle is rather cumbersome to install and somewhat exacting to build. Even more exciting when in operation are random switching devices which are designed to switch the lamps purely at random. Most people are surprised to learn that pure random operations of any device are among the most difficult to achieve in practice and require some very sophisticated techniques. In this section a compromise has been offered in that the switching device does follow a pattern but one which is generally independent of control by the operator.

A modified random sequence programmer

Before describing the construction of the apparatus, it is essential to give a general warning. Electricity at the voltages used for power and light within a

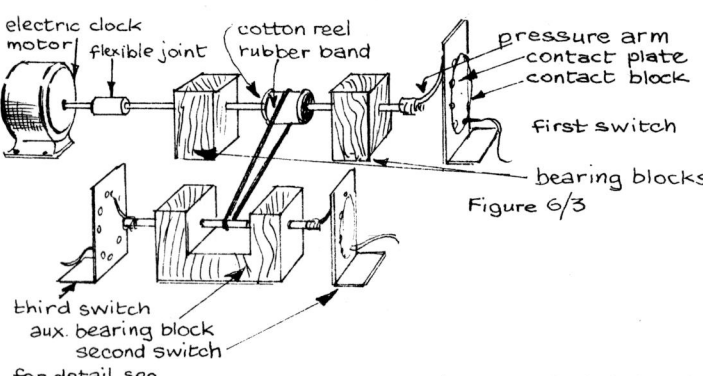

electric clock motor

flexible joint

cotton reel

rubber band

pressure arm
contact plate
contact block

First switch

bearing blocks

Figure 6/3

third switch

aux. bearing block

second switch

for detail see
Figures 6/4 & 6/5

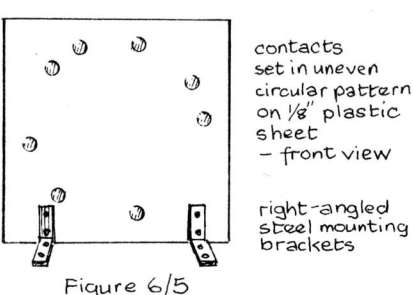

piece of dowelling

pencil clip bent and fastened on dowelling shaft

pencil clip

Figure 6/4

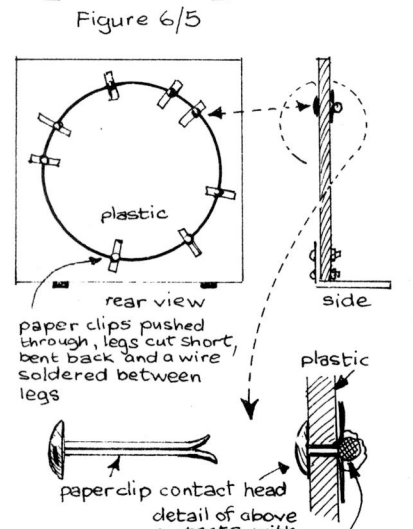

contacts set in uneven circular pattern on 1/8" plastic sheet
– front view

right-angled steel mounting brackets

Figure 6/5

plastic

rear view

side

paper clips pushed through, legs cut short, bent back and a wire soldered between legs

plastic

paperclip contact head

detail of above contacts with wire soldered

home can be lethal. This programmer is designed to switch such supplies and there is no doubt that the open circuits of this device must be guarded by an enclosure which is completely child-proof.

The programmer is constructed as follows:

1. Mount an electric clock motor and extension shaft as shown in figure 6/3. A cotton reel is fixed between the two bearing blocks, but the details are generally as for figure 6/2.

2. Make three pressure arms from the removable type of pencil clip. These are fitted to the shafts where shown in figure 6/3 but construction details are given in figure 6/4.

3. Three contact blocks are made from $\frac{1}{8}$ inch plastic sheet, as shown in figure 6/5. The materials used for the contacts are ordinary brass paperclips. Holes are drilled in a more or less random pattern in a circle in each block, the pattern and number of holes must be different for each. The paperclips are pushed through the holes; the legs cut short and bent over to allow a circle of heavy copper wire to be soldered between each of them. Finally, light steel right-angled mounting brackets are screwed to the block so that it may be mounted in a vertical position. The dimensions are not critical but the circle of contacts must be close to the diameter of the circle which will be described by the ball end of the rotating paperclips.

4. A sheet of springy, shim brass is cut to the same size as the contact block. Shim brass is thin, hard rolled sheet brass. The bottom of this sheet is bent at right angles.

5. An auxiliary bearing block is cut from pinewood to the shape shown in figure 6/3. Again the dimensions are not critical but must be adequate for the purpose served. A short dowel shaft is fitted where shown.

6. Mount all components on a base board, ensuring that each of the three switches fulfils the requirements as given in figure 6/6. It is important that the brass sheet be slightly dished in order that the additional springiness resulting will cause the sheet to spring out of contact as the local force of the pressure arm moves on to another part of the sheet.

7. The rubber band (which must be temporarily put in place before assembly) drives from the cotton reel pulley on the main shaft to the auxiliary shaft. The design is improved and the pattern of switching nearer to random if two auxiliary shafts are fitted, one for each of switches two and three. If this is done, the driving pulley for each auxiliary shaft drive should be of different dimensions, so that the driven shafts rotate at different rates.

When the assembly is completed and each switch is working satisfactorily, a lamp is connected into each of the circuits controlled by each of the switches. Rotating at different rates, the switches will operate the lamps at different times. There will eventually be a pattern established in the sequence of colour combinations, but this pattern will repeat only after many, possibly several hundred revolutions of the clock motor. In this respect the programmer described can be considered as random for all practical applications.

Multi-lamp programmers

By using components from a fairly sophisticated engineering construction set, as sold for hobbyists, it is possible to construct a programmer to the

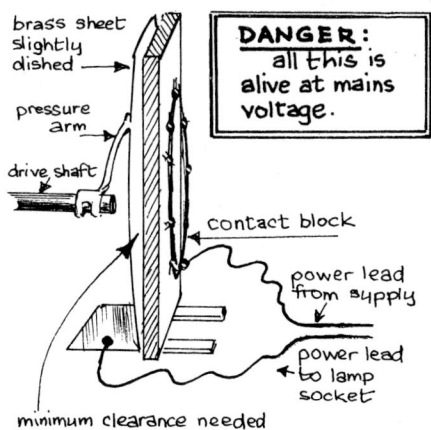

Figure 6/6

minimum clearance needed
to ensure pressure arm forces
brass sheet against contacts
in turn as arm rotates

basic pattern of the one given, but which is capable of controlling a complete bank of lamps. There are two possibilities at least to consider with the use of multi-lamp banks; the individual lamps can be restricted to the three primary light colours so that the colour patterns are formed by the overlap of the lamps; alternatively a range of different filters can be used in conjunction with individual lamps and the colour display thrown over a much wider area.

Regarding lamps most suited to programming there is no absolute criterion for selection, as almost any type and rating can be used. However, there are considerable advantages in using internally reflecting colour flood lamps, such as Philips comptalux E27, which can be obtained already filtered to give red, blue, green, yellow, and normal 3,000:K light.

Applications for programmers

Any craftsman who uses a colour programmer has a highly sophisticated device capable of producing a wondrous display of colour patterns. However, the programmer itself is only a beginning, for much of the art depends on the way in which the apparatus is used. It can be truthfully said, that there are as many possible applications as there are artists to think them up, but that in itself does not absolve this text from making some suggestions.

Rainbows at will

One of the most fascinating displays of light that this writer has seen was ludicrously simple to set up. The three coloured floodlights, connected to the programmer, were held by a partly open window to shine into a garden and on to shrubs and trees. The ephemeral patterns of rainbow colours so displayed, the dancing phantasmagoria caused by the wind movement through the bushes and trees, all had a magnetic dream-like quality. When later, it rained heavily the scene became indescribably beautiful.

Ceilings that glow

A patterned or textured surface on a ceiling can be obliquely lighted in colour patterns. The position of each of the lamps will depend on the depth to which the texture or pattern is sunk. The shallower the texture, the closer the lamps are placed to the ceiling. The more pronounced the texture, the wider apart they can be placed in order to produce a play of coloured shadows. A typical installation of this type is sketched in figure 6/7.

A waterfall of light

A wide alcove or short end wall can be backed with heavy gauge aluminium cooking foil of the quality supplied to caterers. The foil backing will be more effective if it is curved up to extend along the ceiling for a few feet and similarly down along the floor. An efficient installation is shown in figure 6/8. The curved corners between the wall and floor, and the wall and ceiling can be made conveniently by use of expanded polystyrene plastic blocks, cut to an inside curve as seen in figure 6/8. It is found that polyvinyl acetate emulsion cement will hold the aluminium foil and the plastic, without attacking the latter as some cements will do.

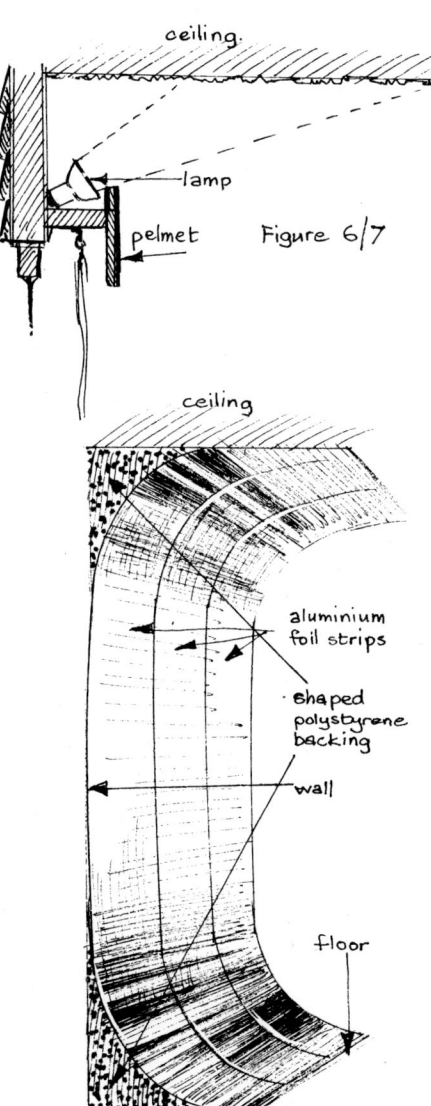

Figure 6/7

Figure 6/8

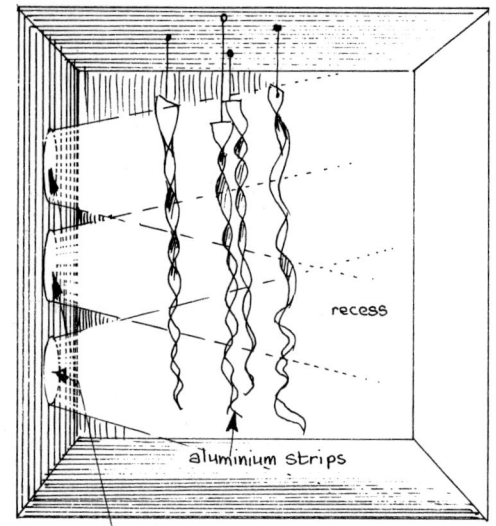

recess

aluminium strips

lamps recessed into wall

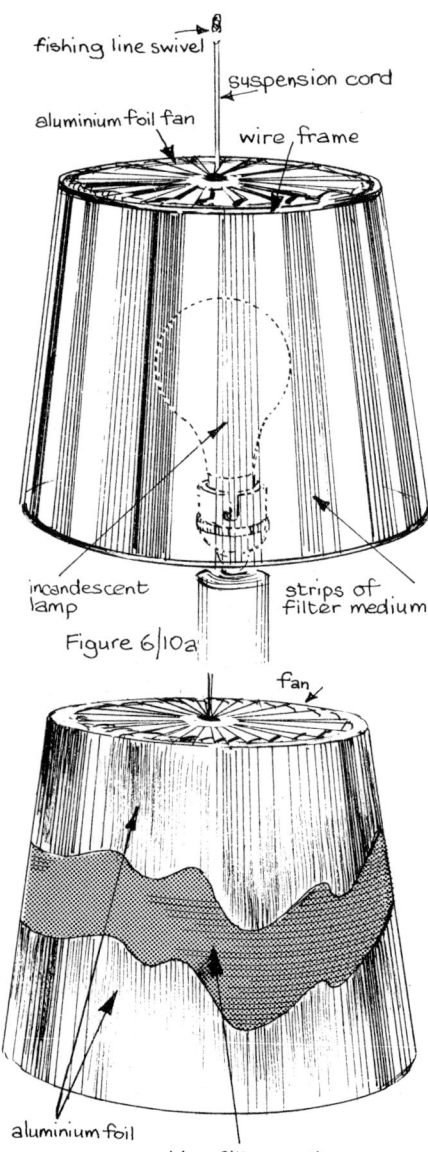

fishing line swivel

suspension cord

aluminium foil fan

wire frame

incandescent lamp

strips of filter medium

Figure 6/10a

fan

aluminium foil

blue filter medium

Figure 6/10b

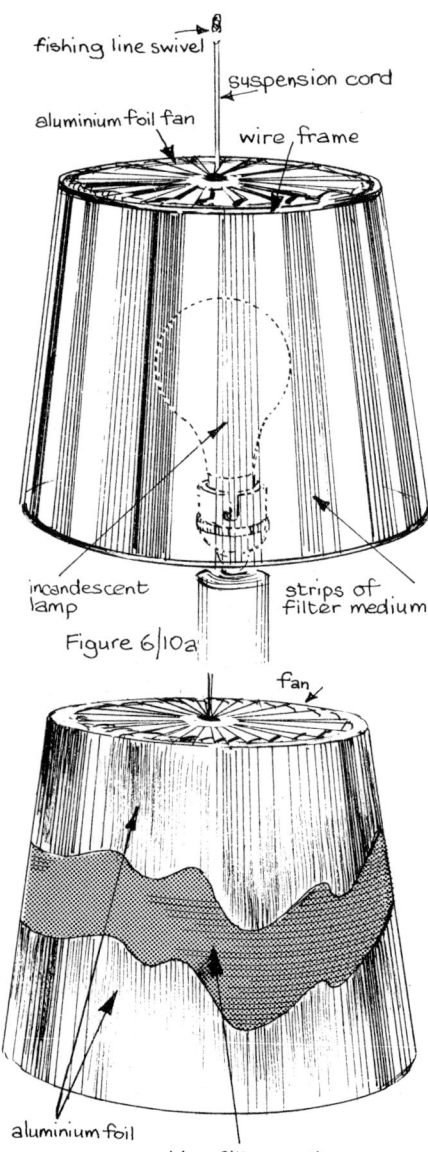

aluminium strips

Figure 6/9

length to suit

When the foil is firmly held in place it can be gently patterned with a series of wide shallow dents produced by striking the surface with the heel of the hand. The lamps are placed close to the ceiling and arranged to shine into the curve and downwards to spread across the wall. As the colour changes take place it is easy to detect a similarity between the flood of light and a waterfall.

Twinkling colours

In the final example given the lamps are sited at the side of the display. The patterns are broken up as well as reflected by a series of aluminium foil strips, arranged as shown in figure 6/9. Note that the strips are twisted in a slow curve and suspended by nylon cords, so they are free to move and spin.

Variations on these basic concepts are possible and many other display screens have their own appeal. For instance, all the examples detailed are front lighted, whereas back lighting on to a translucent screen has much to commend it. There are many possibilities with the use of perforated screens. Foil, pierced with numerous small holes, can act as a multitude of pin-hole lenses with each hole producing a star-like effect. There is really great scope for such entirely unique light compositions.

Variations on a theme

The simplest method of colour changing can produce effects which cannot conveniently be achieved any other way. Consider figure 6/10A. This figure shows a very simple adaptation of the colour wheel, which is rotated by the rising hot air currents of an incandescent lamp. Constructional details should be reasonably clear from the diagrams but the following points are worth special mention.

1.   The fan is made by laying a wire ring of, say, 6 inches in diameter on a sheet of heavy aluminium foil and wrapping the edges of the foil around the ring. See figure 6/11A. With a sharp knife the foil is cut in the pattern shown in figure 6/11C.

2.   A quality fishing line swivel is used in the suspension cord to allow free rotation.

4.   The filter medium is fastened to the top and bottom rings with contact cement.

4.   A 100 watt standard silicon-coated electric incandescent lamp is used both to give light and to provide the air flow which will cause rotation.

5.   A device for producing a fluctuating or wavy coloured line across an object is arranged as shown in figure 6/10B. Such a project was suggested in Chapter 5 under the general heading of *Accent lighting*.

6.   A semicircular aluminium foil reflector can be used to improve the performance of this type of colour project. See figure 6/12.

Random colours

A device which is more amusing than practical but which gives a fully random pattern of coloured lights is shown as figure 6/13. A few explanatory notes are needed to supplement the construction details given.

1.   Broken coloured glass pieces are cemented to rubber or elastic suspension cords with latex cement.

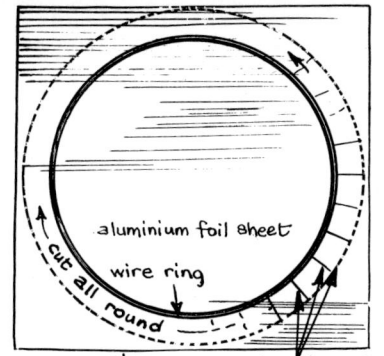

Figure 6/11a

aluminium foil sheet

wire ring

cut all round

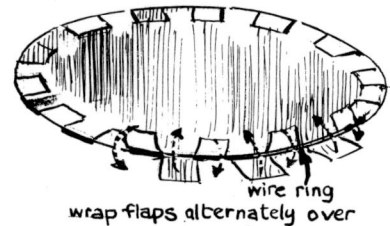

wire ring
wrap flaps alternately over
and under to hold ring in place
Figure 6/11b

cuts in foil cutting foil for wrapping ring.

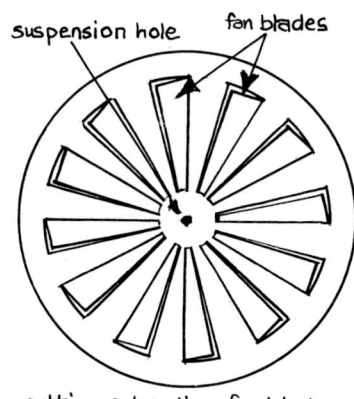

suspension hole   fan blades

cutting & bending fan blades
(all in the same direction)
Figure 6/11c

2.   Each cord is in turn cemented to the underside of a sheet of thin shim steel which is tacked to a pair of wooden sides; the other two sides are left open.

3.   A cheap electric bell has the gong removed and the hammer increased in weight by a ball of modelling wax in which lead shot has been embedded. The bell is in turn cemented on top of the shim steel.

4.   A beam of white light is directed through the open sides and on to a display.

When the bell is switched on the heavy mass of the modified hammer reduces the period of vibration and sets up a series of up-and-down movements in the springy steel plate. Transmitted to the rubber bands, this movement causes the glass pieces to jazz. The light from the lamp is reflected and transmitted through the glass and shows as random colour patterns on the screen. The effectiveness of the moving pattern is dependent, to a large extent, on the number of glass pieces used; if too few, there is too much white light thrown on to the display.

Further colour displays

The number of examples of ingenious ways of creating an illusion of movement in static objects could be extended almost indefinitely so it is felt that there is little point in offering any more. Throughout this book an attempt has been made to impart sufficient basic knowledge to evoke craftsmen (or craftswomen) to undertake planning for themselves. There is, however, one major class of moving display which will now be considered.

Coloured light and water

A display which is always fresh and appealing and which finds a natural place in every outdoor and many indoor settings is the beauty of water and light. Many artistically minded people feel that the splendid attraction of moving water illuminated by coloured light is some compensation for the technological world in which they live.

Be that as it may, there are few who are not enchanted by the tinkling and twinkling of lighted fountains and waterfalls. It is not proposed to delve in great detail into the application of coloured light to water, for most of what has been said earlier is applicable to coloured patterns in or on water.

However, the following brief comments are offered as a guide;

1.   Electricity at household voltages and water is a deadly combination. There is only one rule allowable if electrical fittings have to be used near, or under water, use approved fittings (no compromises) and have the work installed by a thoroughly competent electrician. The one possible exception to this rule is to use a low voltage safety transformer (properly installed) and reduce the voltage to a less lethal level than is used inside the house. Even then it is wiser to avoid any work that is not subject to the most exacting standards.

2.   It is far safer to beam light from a dry distance on to a mirror which then acts as a secondary source of light. Mirrors, after all, do not explode if water drips on them.

3.   Water can be used indoors. Small electrically driven pumps, capable of circulating a suitable quantity of water to a short distance above a miniature

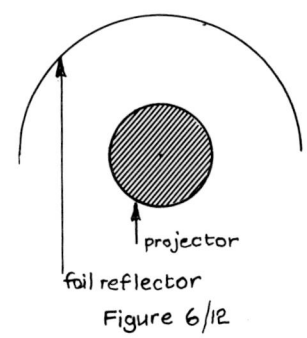

projector

foil reflector

Figure 6/12

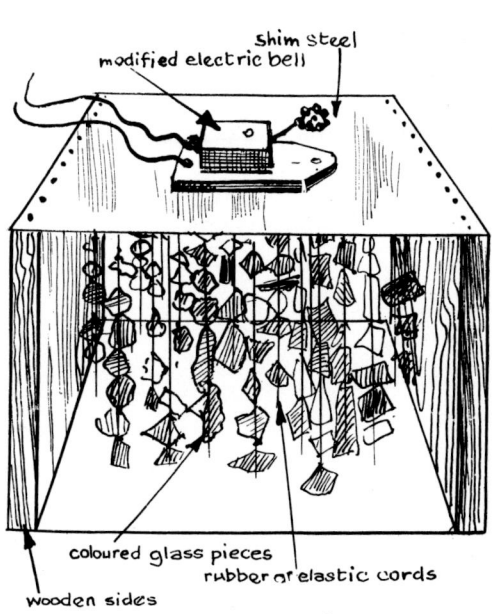

shim steel

modified electric bell

coloured glass pieces
rubber or elastic cords

wooden sides

Figure 6/13

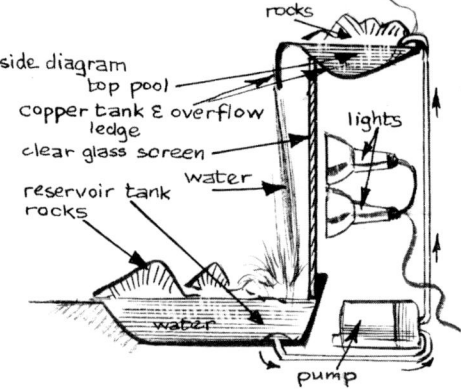

side diagram
top pool
copper tank & overflow ledge
clear glass screen
reservoir tank
rocks
rocks
lights
water
water
pump

front view of miniature
water fall lit from
behind

Figure 6/14

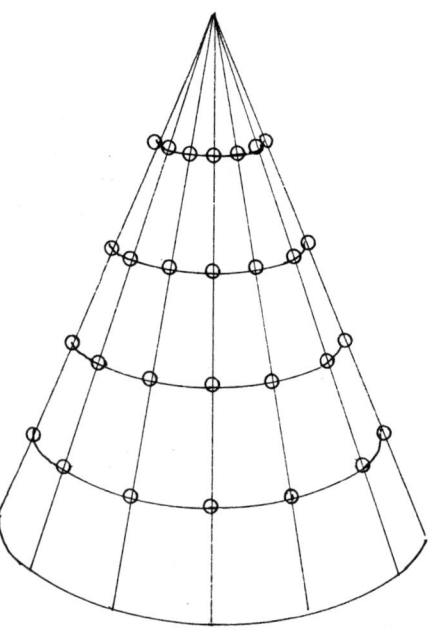

3 Dimensional display with 4 concentric
channels and 12 radial channels (total
48 lamps, 16 control channels)

Figure 6/18

holding pond are highly practical and readily available.
Figure 6/14 demonstrates one possible apparatus.

A few constructional notes may help:

(a)   The miniature waterfall can be sited in a room-divider, a planter, on the side of a fireplace or in some other featured part of a room;

(b)   Copper should be used for the tank and all seams should be brazed;

(c)   The apparent amount of water flowing is greatly increased if made to spread out across a clear glass sheet which is back-lighted;

(d)   The lamps are kept well clear of any possible accidental contact with water having two sealed glass sheets interposed between them and the water.

The light-organ

There are few innovations which have so quickly captured the younger generation (and the young-at-heart in any generation) than the light-organ which turns music into dancing light. A light-organ, in its entirety, is a compound device comprising the following three pieces of equipment.

1.   The source of instruction. This may be a radio, an amplifier used in conjunction with a musical instrument, a record player, or the output from any amplifying equipment which turns sound into electrical signals. A microphone is an example.

2.   The converter, which has as its input electrical signals which vary in accordance with variations in the sound applied to the source of instruction, and, as its output, controlled (modulated) power sufficient to make lamps light to full capacity when required. The converter is the heart of the system and is discussed further below.

3.   The lighting system, which is identical to those discussed earlier.

Modulated light

Light is said to be modulated by sound when a sound signal is so faithfully carried by a beam of light that a faithful rendition of the original sound may be extracted from the signal carried on that beam of light. Sound-organs do not normally reach this standard of perfection but vary according to the care given to the design and construction of the converter.

In this chapter two circuit diagrams, figures 6/15 and 6/16, are guides to building a converter. As with the earlier example of an electronic device in Chapter 3, the circuit diagrams convey ample information to enable a competent electronic technician to faithfully carry out instructions. Whether a craftsman does the construction or has it done is immaterial, for the instructions allow for a completed model to be connected on the input side to any suitable source of instruction.

Two circuit diagrams have been given here, for the standard of performance of each machine differs considerably. This matter is worth discussion. Figure 6/15 is a diagram of a light-organ converter that is comparatively cheap to build, is easily assembled from the components and is generally up to a standard of performance equivalent to most models available on the market. This converter can be built at a much cheaper rate than any proprietary organs offered commercially. Figure 6/16 is a refined version of the first model with quite superior characteristics.

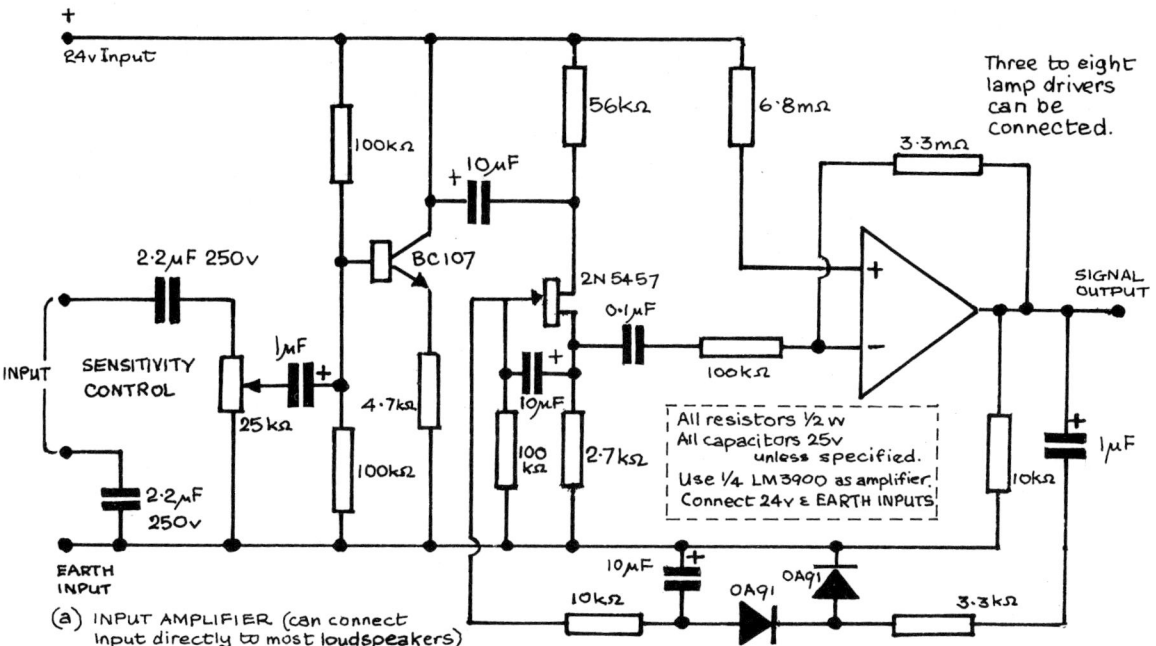

(a) INPUT AMPLIFIER (can connect input directly to most loudspeakers)

LIGHT ORGAN. This figure contains the basic equipment. Up to 8 of the lamp drivers (Figure 6/16) may be added as required.

Figure 6/15a.

Three to eight lamp drivers can be connected.

All resistors ½ w
All capacitors 25v
  unless specified.
Use ¼ LM 3900 as amplifier.
Connect 24v & EARTH INPUTS

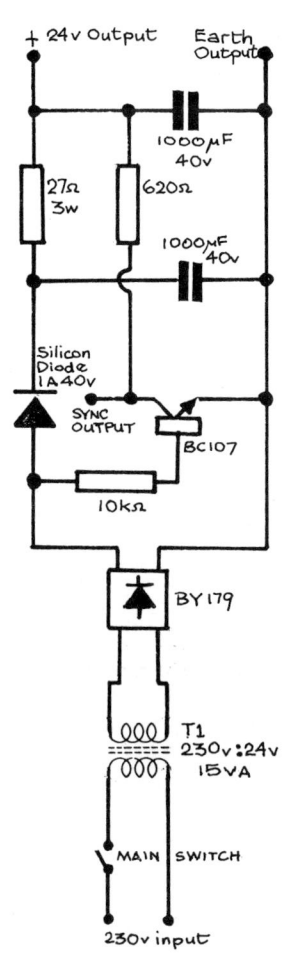

(b) POWER SUPPLY

Figure 6/15b

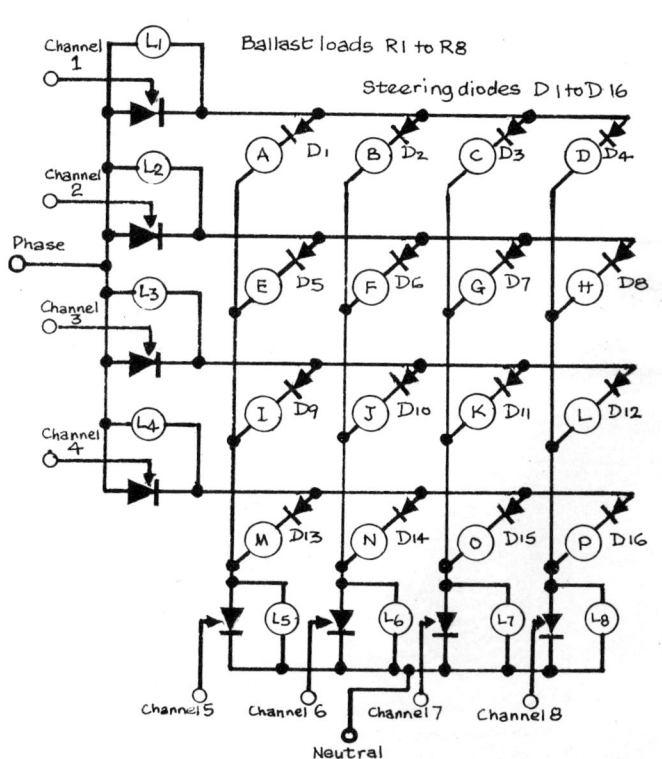

Ballast loads R1 to R8

Steering diodes D1 to D16

A 4×4 matrix where A to P are lamps (60w. each max.) L1 to L8 are 15w bulbs. D1 to D16 are diodes (rated at 600v p.i.v 1A) Thyristors for channels 1 to 8 must each be rated at least 400v 2A.

Figure 6/17.

No. of drivers	CAPACITATORS $C_1 = C_2$							
	UNIT NO. 1	2	3	4	5	6	7	8
3	·1	·022	·0047					
6	·15	·047	·022	·01	·0068	·0033		
8	·22	·1	·047	·022	·01	·0068	·0033	·0022

All resistors ½w. All capacitors 25v.

Figure 6/16   LIGHT ORGAN : LAMP DRIVER

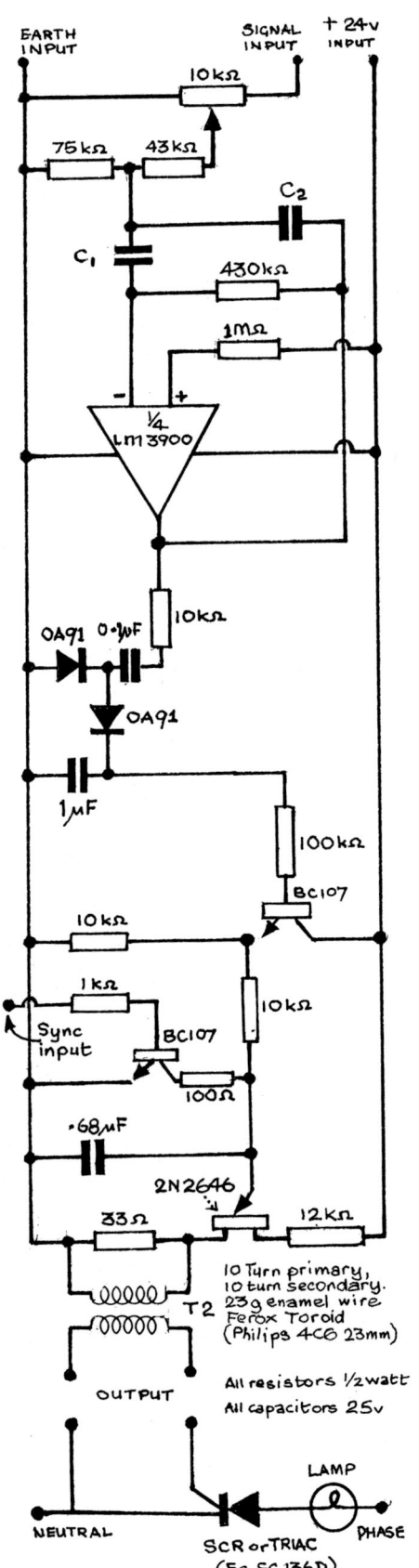

Lighting display

A light-organ splits musical (and other) sounds into several restricted frequency bands, that is base, medium and high sound frequencies if three fundamental bands only are considered. Having so analysed the sound, it feeds energy into several circuits (the circuits equal the number of the frequency bands) in proportion to the volume of sound in the corresponding band. Each circuit contains one or more lamps which will respond with brilliancy equivalent to the amount of energy fed into that circuit. From this it follows that as the volume of sound at a particular frequency increases so will the brilliance of the light corresponding to that sound frequency increase.

In theory, the effect of the correspondence between sound energy levels and the brilliancy of lamps (these may well be of the primary light colours) will lead to an attractive display of light which will change as the music changes. To a limited extent this is true, but people do not respond favourably to sudden changes in illumination levels especially rapid changes. Much more pleasing to the eye are variations in patterns of light, accompanied, preferably, by equivalent differences in colour, and assisted by a not-too-rapid change in the intensity of the light. This is the main advantage of following the more complicated circuit diagram (figure 6/16), rather than the simpler one (figure 6/15).

Matrix display

The most sophisticated development in the use of a light-organ, namely, to provide energy for a series of separate circuits in accordance with variations of sound input, is the arrangement of several lamps in a matrix. The theory of matrices upon which these advanced light displays are based is complex, but the circuit shown in figure 6/16 is designed to make use of this theoretical complexity in producing displays which form unique light patterns.

A quite fascinating matrix display can be accomplished by using an 8-channel organ and a 16-lamp matrix is far overshadowed by a 48-lamp matrix controlled by 16 channels. The capital outlay of a 48-lamp matrix with the necessary control-organ is considerable. However, the patterns of ever-changing loops, whirls and circles are fascinating.

Figure 6/17 shows the connections for a 16-lamp matrix as connected to an 8-channel organ, and figure 6/18 is the equivalent, but more complicated circuit for a 48-lamp matrix used with a 16-channel control.

These matrix displays are in their elementary stages, and there is tremendous scope for quite simple variations, such as (a) the placings of lamps, (b) the colour disposition of the filters associated with individual lamps, (c) the actual settings of the controls on the colour organ and (d) the choice of music fed into the organ. In effect, this combination of colour organ and matrix display is very much an experimental art form, but one which shows promise of being the basis of some creative art of the late 20th century.

This book has covered a craft which began as a crude utilitarian necessity in the dim dawn of man's history and which has advanced to the point where it has become the most sophisticated technologically, of all art forms. The story is unfinished, for the new revolution with lighting as a creative art is just beginning. Where it will end no one yet knows but there is infinite pleasure to be gained by participating in this revolution.

Experimental lighting modes

There are many developments in that fringe area between pure electronics and lighting applications which will profoundly change the available sources of light within the next decade or so. Some of these, such as the laser and the holographs produced by laser light, are spectacular and have been mentioned in earlier chapters. Other applications involving somewhat obscure theory are not well known to artists although some of the effects achieved are capable of displaying delightful colours and delicate patterns not obtainable in other media.

This chapter will briefly touch upon a number of possible modes in lighting. The text is technically based but simply explained in order to show that no technological background is required in order that the underlying principles can be applied in creative artistry. No attempt has been made to give all necessary details for the construction of apparatus needed to demonstrate all the effects but it is hoped that sufficient has been said to allow those whose imagination is captured to pursue the matter further.

Spectral colours
There is no cleaner, brighter, richer source of colour than that which appears when white light is split into the rainbow hues of the visible spectrum. As was earlier discussed, a glass prism can effectively separate white light into the several wavelengths of the spectrum but does not accomplish this as effectively and dramatically as can be done with a diffraction grating. *Diffraction gratings*, as shown diagrammatically in figure 7/1, consist of a high grade optical glass sheet on which has been scribed a number of equidistant grooves. In optical work (in the visible spectrum) as many as 10,000 to 30,000 separate lines are scribed to each inch of width, a diamond point mounted in a precise mechanical device known as a 'ruling engine' carrying out this task automatically. Light passing through the narrow widths of clear glass between the grooves is subject to a phenomenon known as diffraction. Any standard text, dealing with the physics of light, discusses this subject, possibly under the heading of 'fraunhofer' and 'Fresnel' diffraction. The theory of diffraction through a slit is a little complex but does not need to be known in order to apply the results.

The type of grating shown diagrammatically in figure 7/2 is known as a transmission grating owing to its action depending on light being passed through the grating. Similar gratings ruled on metal sheet are reflection gratings for obvious reasons. The diagram is showing the bending of light rays from the action of the narrow transmission slit. For the sake of clarity only one wavelength of light (monochromatic or one coloured light) is illustrated there. The theory of diffraction shows that each wavelength of light (that is each colour in the spectrum) is subject to its own particular angular deviation whereby it is bent through angles A and B as shown in the figure. For instance, the angular deviation of violet light is 13° 40′ (said to be thirteen degrees, forty minutes of angle). From the opposite end of the spectrum, red light is bent with an angular deviation of 24° 30′.

From the differences in angular deviation of lights of the various colours of the spectrum comes the ability of a diffraction grating to separate out

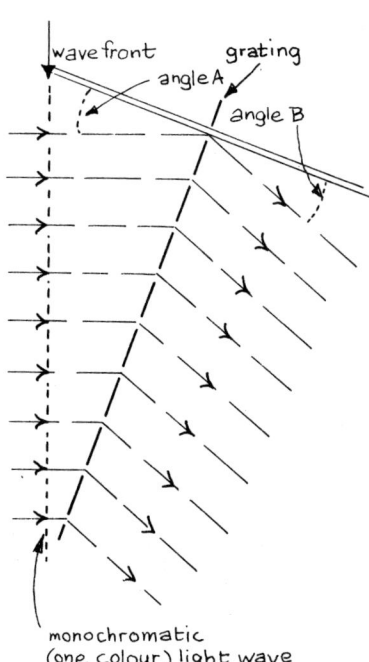

monochromatic
(one colour) light wave

Figure 7/1
action of diffraction
grating.

the parallel rays of white light, incident (falling upon) on the grating, into the various component colours of which the white light was originally composed. This is shown for two wavelengths as figure 7/2. It is pertinent to note that in a prism the violet wavelengths are bent most and the red least, whereas with a diffraction grating the opposite is the case.

For the purpose of art work the great advantage of a diffraction grating over a prism for casting a spectrum lies in the simplicity and low cost of the ancillary equipment needed for projecting an effective spectrum, figure 7/3. Photographically reproduced diffraction gratings of quite large sizes can be obtained from dealers in scientific apparatus. Such gratings are economical in cost and fully effective for anything but the most exacting applications. These photographic replicas can be ordered as parallel linear gratings, where the lines are straight ruled or as parallel circular gratings where the original grooves from which the photographic positive was taken, are concentric circles. More novel effects are also offered where the lines are non-parallel, two or more sets of lines cross, etc. These throw various kinds of patterns in colours and bands.

Any common form of projector lamp can be used without alteration for producing spectacular colours from diffraction gratings. Spectrum effects can be used to great advantage to enhance plain backgrounds, thrown against the so-called 'specular' screens as used for home movie projectors and in other situations where the purity of colour obtained can give results not obtainable in any other way. In consideration of the inexpensive equipment required, there is merit in experimenting with a medium size photographic reproduction of a diffraction grating for personal assessment of the effects which can so readily be effected.

Polarised light
Light in the visible spectrum is a narrow band of the very much larger spectrum of electro-magnetic waves. This spectrum is shown in brief detail in figure 7/4. In this figure is shown a continuous series of spectra with an enormous range of wavelengths by which energy is radiated into space at different electro-magnetic frequencies. At the lower end, the slow frequency power currents of 50 to 60 or higher pertz (one pertz is one complete cycle per second as shown in figure 7/5) have a wavelength of more than a million metres. At the other end are the very high frequency energies of the gamma rays emitted from atomic nuclei and the even more energetic cosmic rays from outer space. At this end of the spectrum are wavelengths of 10–15 metres and frequencies of 10^{23} pertz.

Lying about half the distance along the total electro-magnetic spectrum is the band of radiation to which the human eye is sensitive. This is the optical or visible spectrum comprising one octave only of the whole, with violet light 3.9×10^{-7} metres to red light, 7.8×10^{-7} metres. Red wavelength is twice as long as violet. Below the red lines lies infra-red (heat) radiations and above the violet is ultra-violet with wavelengths from 3.9×10^{-7} metres upward to well into 10^{-8} metres. Ultra-violet radiation will be discussed at length later.

Visible light is therefore only a narrow band of the same class of phenomena which provide radio, television, radar, X-rays and other such energies. Visible light is emitted from a transmission station in much the same way as a commercial radio advertisement. With visible light, the transmission begins in the outer electrons of atoms which absorb energy either as heat energy, electrical energy or in one or other of several ways and then this extra energy is emitted in the form of small bursts of electro-magnetic waves, known as 'photons'. These photons travel outward from their source in all directions and the miniature wave-trains of energy (which are the photons) vibrate in every possible plane. It is difficult to show this in a two-dimensional drawing but figure 7/6 attempts to do so.

Under certain circumstances part of the light emitted is blocked, as for instance, when passing through a polarizing slit as seen in figure 7/7.

Figure 7/2
Angular deflection of red and violet light through diffraction grating.

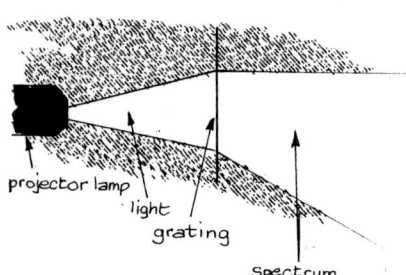

Figure 7/3. Producing spectrum with diffraction grating

More commonly, unpolarized light striking a reflective surface at a critical angle (known as 'Brewster's angle' after the man who first investigated the phenomenon) is reflected as plane polarized light which is parallel to the plane of the surface. For most window glass, Brewster's angle is 57°. This effect is shown in figure 7/8. Glass is not the only reflective surface, which under the correct circumstances can cause plane polarization of light. Water, both in large bodies and as water films on roads, behaves in a similar fashion.

Several naturally occurring minerals exhibit an ability to polarize light. The best known of these is tourmaline, a complex silica mineral. Iceland Spar, a form of the mineral calcite, is another example. These two and other crystalline substances apparently have a micro-structure aligned in one direction so that only plane polarized light will pass in a given direction. In the early 1930's H. E. Land marketed a synthetic polarizing material consisting of a transparent artificial resin in which were suspended micro-crystals of a substance, herapathile. This combination acted as an effective polarizer. Other types of artificial polarizing materials are available today in addition to the 'polaroid of Land's'.

If an arrangement as shown in figure 7/9 is used, most transparent specimens when subject to stress (compressive stress from a weight or tensile stress from pulling) will exhibit patterns similar to the oversimplified form indicated in figure 7/10. The patterns can be caused to change by variations in the amount of stress applied to the specimen and multi-coloured, ever-changing patterns can be produced by quite simple mechanical means, a selection of which are offered in a companion book in this series 'Kinetics'. An easy way to observe the fascination of polarized light is to try out an arrangement, such as in figure 7/9, with two pairs of Polaroid sunglasses acting as screens, an electric torch as the source of light and a crumpled ball of cellophane as the specimen.

Lasers and holography

For artists with well-lined pockets or a knowledge of technology which must be somewhat more advanced than that of the ordinary home mechanic, there is opened a field of visual communication which is as exciting as it is new. It is not proposed to go deeply into this subject here but it is relevant to give sufficient details to allow those who may see potential in this medium to judge whether the complications of the techniques required are worth pursuing further.

Rather more important (from the point of view of an artist) than how a laser operates or even how laser light differs from ordinary light, is an elementary idea of the effects of laser lights and these points will be discussed first. Of the greatest significance is the fact that laser light is dangerous. Even low energy lasers emit illumination many times more damaging to the eye than even noonday sun. This is an aspect of laser operation which has been well publicised but it must always be a limiting factor in laser operation. With the above in mind it is possible to give thought to some details of laser operation.

There are a wide selection of lasers offered commercially at present and more being developed. The types with appeal to artists will be those producing light in the visible spectrum. The light will be monochromatic (one coloured) and polarized to an extent not known before the development of these devices. The laser effect has been likened to that of a massed regiment marching in step over a swing bridge and producing massive movements in the bridge owing to the synchronised weight of each soldier falling on the bridge at each step. The bridge under these circumstances, will resonate at its natural frequency, building up gross surges of movement to the point of destruction and beyond. Ordinary light as emitted from a domestic lamp is similarly described as being akin to the same soldiers marching in broken step so that the phenomenon of resonance does not occur.

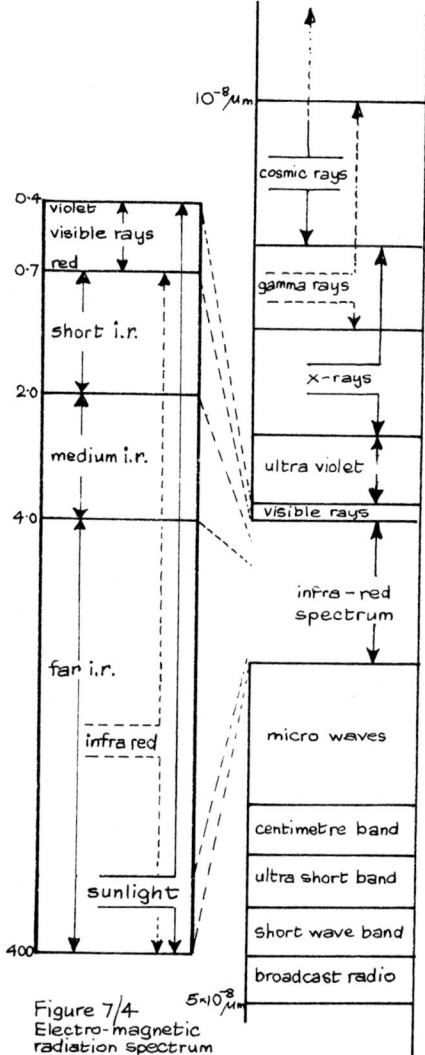

Figure 7/4
Electro-magnetic
radiation spectrum

From the above it follows that light from a laser is known as 'coherent' light with every wave front being in step. From the point of view of artistic expression, coherent light has several interesting characteristics. These are worth tabulating as follows:

1. The intensity is several magnitudes greater than ordinary light and while this is a dangerous property as already considered, it can, if used with caution, be the source of intense flashes of light in, say, a kinetic construction which will show up even in direct sunlight. To so use coherent light without danger to viewers' sight, the light must not be reflected from highly polished surfaces but directed against specular reflectors (matt surfaces) which will break up the beam. A spectacularly successful way of utilising this intense, beam-like property of laser light is shown in figure 7/11. The tank can be made from glass-sheet, cemented with epoxy resin as described in the companion book 'Glass Crafts'. The mirrors can be fastened in place in a like manner. A fair choice of materials to be formed into a 'gel' with water are available—dextrin, gelatin and starch mixed in warm water are found in most homes.

Other ways in which laser light can be used will result from adaptations of some of the suggestions made earlier. No apology is offered for once again stressing the danger of coherent light with its inherent probability of permanently damaging the retina of any eye upon which the unchanged light may fall.

2. Laser light is a pure colour. In this respect it is far more confined in the band of radiation emitted than even the best possible filter. There is little possibility of changing the wavelength of the colour produced by fractional lasers so that choosing an effective colour is important in the purchasing or making of the device. It may be mentioned that monochromatic light cannot be altered in colour by passing through a filter or reflecting from a surface.

3. The difference between ordinary light and coherent light is no more strikingly demonstrated than in the effect obtained when an ordinary surface is illuminated by laser light. The coherence of the light from a laser causes the wave fronts, reflected from the many tiny imperfections of even the smoothest surface, to interfere with one another. Where the waves overlap, figure 7/12, separate waves reinforce each other and the result is a bright spot. Where waves cancel, figure 7/13, then there is a lack of light at those places. This interference pattern is characteristic of coherent light but not of ordinary light because the light waves originating in the laser are all in 'phase', rising and falling together. This does not happen with other light radiation.

The minute imperfections of the surface viewed reflect coherent light at slightly different angles so setting-up interference patterns. Every movement of the head or eyes subtly changes the line of sight with which the surface is seen and causes a scintillating series of sparkles and miniature simulated explosions. This is a phenomenon with an appeal all its own and one that can be recommended as well worth pursuing.

The holograph

Of all the uses that may be made of laser light sources in the field of art, the most powerful, the most striking and the one which will undoubtedly be of great significance is the application of a surprising technique known as holography.

A holograph, as viewed, is an almost unbelievable three-dimensional reconstruction in light of an object. It has been variously described as viewing through a window, an object swimming in space, a magical representation of the essence of an object, a visual study frozen in time and in many even more fanciful terms. The unbelievable fact is that all the above and more descriptions are the truth. Remembering that related technique is still in its infancy and that every day sees some advance in the sophistication of application it is obvious that scientists have a new

Figure 7/5　One cycle of
Alternating current
or voltage

Figure 7/6
Non-Polarized light
travels in every direction

tool with which to study fundamental optics and artists have an advancing field of visual presentation with which they must become familiar.

A holograph exhibits all the characteristics of depth as well as width and height. The image appears in full three dimensions, not as in painting or ordinary photography as a matter of perspective, but with all the effects of parallax as well. Parallax is that attribute of a solid object whereby background details obscured by closer features move to one side when the head position is changed. A holograph moreover has true depth. In other words not only does the image occupy space up and down (as in a flat photograph) but it also is present backwards in depth. The eye has to refocus between foreground and background and a camera can be brought to focus similarly, to take conventional photographs of the holographic image highlighting background or foreground at will.

The principle of producing a 'hologram', the pattern imposed on film from whence a holograph is reconstructed, is schematically shown as figure 7/14. The action is as follows:

(a)   A coherent beam of light is produced by a laser. This light is such that every wave of light is the identical wavelength and completely in step with every other wave.

(b)   The beam is intercepted by a 'beam splitter', a half-silvered mirror which reflects half the light in the beam and allows the other half to pass through.

(c)   By utilising a train of mirrors the outputs from the beam splitter are made parallel but with the light still fully coherent, that is in step and of the same wave length.

(d)   One of the half-beams, that called the reference beam, is reflected again by a mirror and back on to the emulsion of a photographic plate.

(e)   The second half-beam is similarly reflected back on to the photo plate but this from the object to be recorded.

(f)   The light which is received on the plate is still of the same wavelength but has interference patterns imposed owing to the different lengths of the paths travelled by the half-beams. The situation is similar to that described earlier under 3 above. Where coincidence occurs between the reference wave and the reflected wave there will be reinforcement of the light intensity and the photo emulsion will be exposed at that point. Where cancellation occurs the emulsion will not be exposed.

(g)   On developing, the result will be a hologram which is in no way recognisable as a photograph of the object. On the contrary the pattern on the hologram is a meaningless complexity of fine lines which in total make up a highly complex, most sophisticated diffraction grating, similar to but far more complicated than the gratings described before.

There is nothing unduly complicated in the production of holograms. The possession of a laser is imperative and it must be of the kind known as a 'single mode' laser meaning that all the output is of a single wavelength. The photographic plate must be of a very special type to record the ultra-fine structure of the interference pattern. The plates are expensive but can be obtained. Of more immediate moment is the need for absolute stability of the apparatus. Hyper-fine emulsion, photographic plates are comparatively insensitive, a typical example having an ASA speed of 0.003 in comparison to a normal black and white film ASA speed of 400 or 500. These require long exposure times. With a small laser used the needed exposure can require 5 to 10 minutes to carry out.

An effective hologram will require 60,000 or more separate lines to be recorded clearly for each linear inch of length and breadth of the hologram. The absolute requirement for rigidity in the apparatus is for movements of less than a few millionths of an inch. This is a requirement most difficult to achieve and even more nearly impossible to test for effectiveness of the precautions taken. However, it can be done by using massive blocks of concrete ($\frac{1}{2}$ ton or more) insulated from other structures by pads of foamed plastic or like material so that ground tremors and other

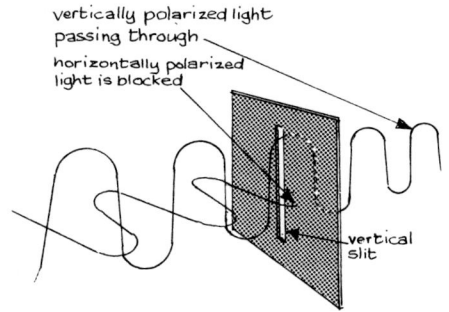

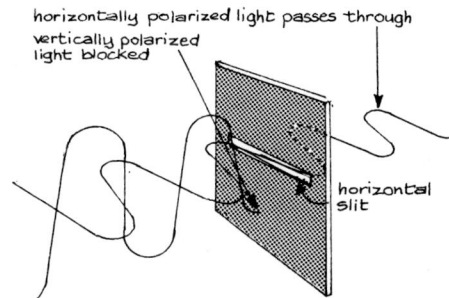

Figure 7/7   Polarization of light

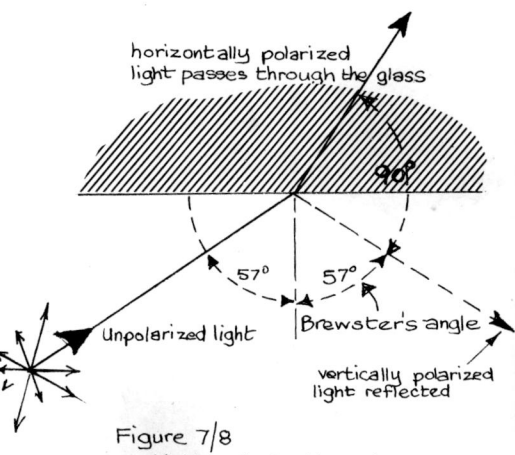

Figure 7/8
Light polarization at
Brewster's angle.

vibrations are damped out. Figure 7/15 gives some idea of the very heavy mounting essential for hologram apparatus.

There are a number of amateur experimenters in optics with facilities for producing holograms and there are many commercial establishments likewise equipped. It should be possible to have holograms produced of a specific piece of work without necessarily having to construct one's own apparatus. Even at this early stage in the development of holography there is the advantage of being able to hold a fully detailed three-dimensional image of pieces which may be ephemeral in character for some other reason not likely to be readily available in the future. For those folk who wish to experiment with holographic images, holograms can be purchased from science supply houses and can be converted into holographs in terms now given.

It will be recalled that a hologram was earlier referred to as a complex and sophisticated, photographically produced, diffraction grating. As discussed before, a diffraction grating bends and disperses light which is passed through it, figure 7/1. Similarly a hologram bends and disperses light which is transmitted through but in this case the dispersion is in a complicated pattern. If light from a laser is used as in figure 7/16 a three-dimensional image is seen by the eye. For a first step it is not even necessary to use a laser to produce an image from a hologram. In a strongly lighted 35 mm slide projector place a good quality light filter of any convenient colour in the slide holder. Fasten aluminium foil across the projector lens and while in place pierce the foil back against the glass with a fine needle point to make a minute pin hole. The thin beam of light from this source can be used in place of laser light for producing holographs as the light from a pin hole is effectively coherent.

This discussion has barely touched upon the subject of holography. In the near future this is going to be a topic of much moment and one which a progressive artist cannot avoid. It would be rash to say that the visual art of the future is to be based on holographs but it would be even rasher to say that this is not a likely trend. The point to be made lies in a suggestion that all people interested in the visual arts be aware of progress in this field and learn something about the art of holography at first hand.

Fibre optics
Yet another comparatively late development in technology has provided artists and craftsmen with an exciting new medium. The principle of using transparent substances to guide light around corners is not new. However, the freely available optical fibres which allow light to be piped along a thin fibre, passing around corners and even complicated knots to emerge at the end of the fibre, are the result of current work in the glass and plastics industry.

An optical fibre in itself is a simple device. The principle upon which the light is confined is even simpler. Refer to figure 7/17. There is shown the path of a single ray of light along an optical fibre. The angles shown in the figure are important although they refer to only one particular type of plastic product and have no significance for glass. The phenomenon upon which light is trapped within a fibre is called 'total internal reflection'. Consider now figure 7/17. The main points to be made are:
(a) No light will enter the fibre unless it falls within the 64° angle cone shown on the left of the diagram.
(b) The maximum angle of incidence of a ray of light on to the end of the fibre is 32° from the line of the axis of the fibre.
(c) With an angle of incidence of 32° the ray will be refracted to an angle of 21° inside the fibre (being bent 11° at the end surface boundary) to strike the inner long surface at 69° and be totally reflected in a line down the length of the fibre. Rays travelling parallel to the axis will travel the shortest path down the length of the light guide.

With the basic angles as given in figure 7/17 to act as a reference it is

1st Polarizing screen

2nd Polarizing screen

weight

light projector

transparent material under stress

Figure 7/9 Producing stress patterns under polarized light

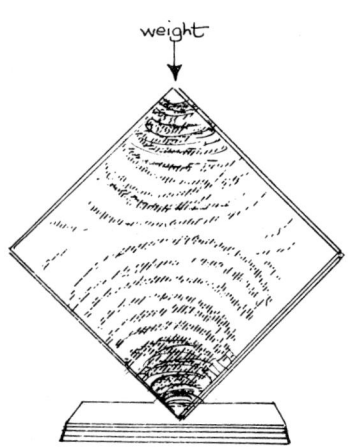

weight

Figure 7/10
Stress patterns in
transparent sheet.

possible to construct a series of drawings to see what happens when light rays strike the end at some other angle than 32°. Figure 7/18 shows this happening at a greater angle. The rays of light do not pass through the boundary but, so to speak, glance off the surface. For this reason this angle of 32° is known as the critical angle of the particular substance under discussion here. Other materials will have different values of critical angles. Figure 7/19 shows what happens when a light ray enters the end of the light guide at such an angle that it is below the critical angle. In this case the light ray is subject to multiple reflections off the walls of the guide but is still confined to the guide. This is the art of the optical fibre maker, ensuring that the light is confined within the fibre. The light ray can be caused to zigzag by successive reflections as many as 500 times in each foot of length of the guide. This represents a very great increase in the distance travelled by the light. For instance a 10 mil (0.010 inch) diameter fibre, which is the most common used, will, if the light zigzags 500 times per foot, travel something over four inches further per foot in taking this path than when travelling one foot parallel to the axis, as shown in figure 7/20.

From the above considerations it becomes clear that the efficiency of the system is dependent on the angle of incidence of the light. The nearer to parallel to the normal of the end surface the incident light is caused to be, the more light will reach the end of the guide. For this reason, it is always advisable to use an optical lens with the light source in an endeavour to direct light on to the ends of the optical fibres. Even a cheap moulded plastic lens costing less than a dollar will increase the amount of available light by as much as 20 times.

Losses in optical fibres. With the particular example under examination here there is a loss of about 9% of the light in the fibre for every foot of length. This does not mean that no light will be seen at the end of a little over 10 feet length of fibre for the decrease is expotential. That is beginning at the source end 9% of the total will be lost in the first foot lengths leaving 91%. The next foot of travel will lose 9% of 91% or 8.19% leaving 82.81% of the original light. The third foot length will thus lose 9% of 82.81% being 7.45% leaving 75.36% of the original light and so on. The practical outcome of this expotential loss is that light can still be seen 30 feet or more from the end but for reasonable brightness the lengths should not exceed 10 feet and from the calculations used as an example above, it is probably wise to limit the length to 3 or 4 feet for visual displays where external illumination is present.

A significant statement was made earlier to the effect that the loss was calculated on the total amount of light which had entered the guide. All the incident light falling on the end will not enter the guide. The transparency or light transmission powers of the acrylic based plastics (in the example studied here the material is polymethyl methacrylate) is as good, if not better, than high quality optical glass. However, even this highly transparent substance causes a theoretical loss of 4% of the incident light when entering the guide and the same value when leaving the guide. Thus a combined loss of 8%. In practice, cutting the fibre ends with a sharp blade (a new razor blade for instance) will increase this end loss to at least 10% and possibly as much as 15%. A combined end loss of from ⅕th to nearly ⅓rd of the available light. To come close to the theoretical value of 4% end polishing must be used. To do this satisfactorily:

1. Gather the bundle of fibres together and after arranging them to lie as parallel as possible coat the ends with a solvent-less epoxy resin. Most proprietary brands of epoxy cement do not include a solvent which will attack the plastic of the fibre.
2. When the epoxy has set, cut at right angles to the axis of the fibres with a fine tooth saw.
3. Polish the cut end in turn with 200 grit, 400 grit and 600 grit abrasive paper or cloth. Silicon carbide paper (the so-called wet and dry paper) is

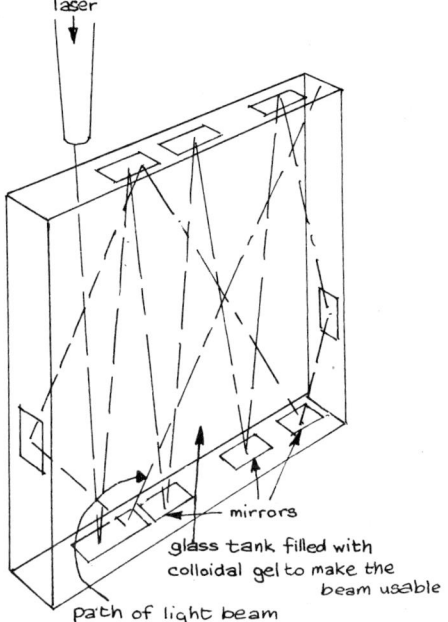
Figure 7/11 Use of laser beam

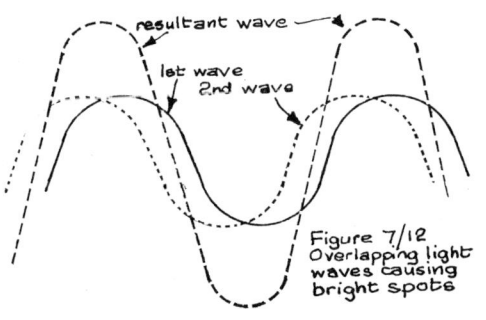
Figure 7/12 Overlapping light waves causing bright spots

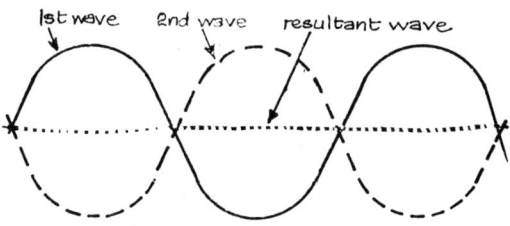
Figure 7/13 Overlapping light waves causing dark spots.

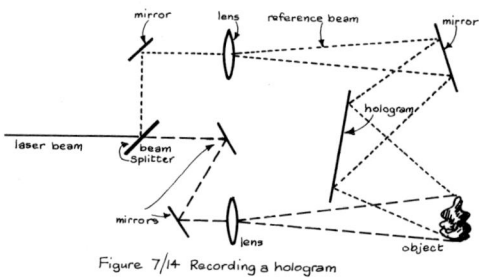

Figure 7/14 Recording a hologram

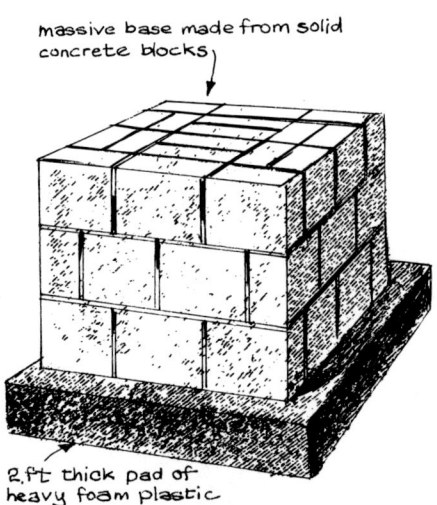

massive base made from solid concrete blocks

2 ft thick pad of heavy foam plastic

Figure 7/15 Stabilizing base for hologram apparatus

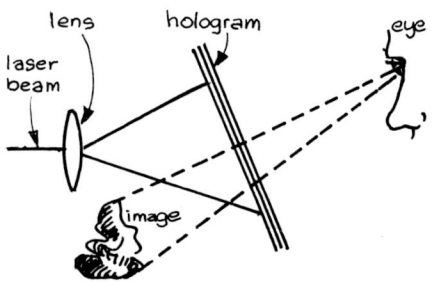

Figure 7/16 Reconstruction of a hologram

near ideal for hand polishing. Finish with a good grade of polishing compound (tin oxide compound is fine).

Alternatively, the cut ends can be freed of a certain amount of roughness at any stage of polishing by applying an acrylic based lacquer. This can well be carried out by dipping. If the lacquer requires thinning (a very thin coat is required) use 1, 1, 2-trichloroethane (toxic) liquid as a thinners. To obtain coloured light output use a coloured acrylic base lacquer but the colour will be obtained at the expense of extra light losses.

Plastic fibres, as used as light guides, must not be subjected to greater heat than 80°C. In a confined space and close to a lamp, this can easily be exceeded. Precautions must be taken to ensure that the design of apparatus is such that this does not occur.

Light can easily be transmitted around quite sharp curves but, as can be seen from figure 7/21, there is an increasing loss of light as the radius of curvature decreases. A couple of factors add to these losses:

1. Reflections from a curved surface can be visualised accurately as being the action of small flat reflecting surfaces considered to be tangentially placed at each point of curvature. From the figure it can be seen that when this is done diagrammatically with due allowance made for the mechanical distortion of the fibre on bending, the light waves no longer proceed in an orderly fashion along the guide, being irregularly reflected from surface to surface. Indeed, when the radius of curvature becomes overly small (the bend is a tight one as would happen if the fibre were knotted) some of the rays strike the wall of the guide at less than the critical angle, enter the sheath and are then absorbed.

2. On the inside of the bend the inner surface of the fibre will crumple and cause even greater scattering of the rays and consequent loss. Generally speaking, it is unwise to subject light guides to tight bends unless some particular effect is required. The human eye is a poor judge of light intensities and large losses can occur in a light guide without a qualitative loss being noticed excepting as an overall loss of brilliance in the display. *Sheath losses and special effects.* The art of the optical fibre maker is to sheath one transparent material in another transparent material so that light is 'piped' down the centre material. The general construction of a fibre can be seen in figure 7/17 and in figure 7/22 where the action of the sheathing plastic in conjunction with the cone, is shown. Reflection, as generally understood, does not meet the facts. Light does not bounce or glance off a reflective surface as a rubber ball bounces off a wall. Rather the light enters into the body of the reflector for the distance of a few wavelengths of light and is then re-emitted from the surface. This is shown, much exaggerated, in the figure.

If the sheathing of the light guide in question is other than transparent the light wave is absorbed. By the same token, if the sheathing is abraded or damaged in any way the light is not returned to the guide. Any practical light guide contains some impurities in its substance and some irregularities in its structure. These cause light scattering which in turn causes the whole fibre to glow with lost light. The amount of loss can be grossly increased by abrading the sheath surface. If the abrasion is accidental the result detracts from the required effect of the fibre use. However, in many cases highly attractive results can be achieved by controlled abrasion (very lightly). There is left a luminous and astonishingly flexible thin column which can be incorporated into a number of compositions including woven and other fabrics.

Without question, optical fibres can be used in a wide range of decorative and utilitarian applications. Provided the not-stringent limitations touched upon here are taken into account, these low cost light guides have a brilliant future, limited only by the imagination of craftsmen and artists.

Electro-luminescence
In late years much of the theory regarding the emission of light from

substances has changed. As a result of greater knowledge of the mechanism of light, production has come about of a number of new substances, entirely man-made, which can become luminous. One such intriguing material exhibits the property described as 'electro-luminescence'.

An explanation of the mechanism of light production is simple and adds so much to the understanding of contemporary developments in electro-optics that it is deemed profitable at this point to consider this subject. Reference to figure 7/23 shows one means of displaying in diagrammatic form the various energy levels which can be explained as existing in the structure of an atom. Several points must be made covering an explanation of the energy level diagram as follows:

(a)   It is generally conceded that a very simplified picture of the structure of an atom shows a central nucleus bearing one or more positive electrical charges and clouds of light weight (in comparison to the elementary constituents of the nucleus) negatively charged electrons forming a nebulous but recognisable pattern outside the nucleus.

(b)   The electron clouds are distributed at regular distances from the nucleus, being traditionally referred to as being in shells known as *k*, *l*, *m*, *n*, etc., in accordance with a coded system beginning at the nucleus and moving outwards.

(c)   The forces which bind the atom together (the electrons to the nucleus) vary inversely as the square of the distance an individual electron is from the nucleus. Hence one electron at a given distance from the nucleus will have 4 times the attraction towards the nucleus as another electron twice as far away ($2^2 = 4$). Similarly, 3 times the distance $\frac{1}{9}$th the attraction.

(d)   From (c) above it follows that the *k* shell electrons are more tightly bound to the nucleus than the *l* shell electrons which are in turn subject to more attraction than electrons in *m* shell, and so on. It therefore follows that each shell position in the atomic structure and in figure 7/23 represents a certain energy level for those electrons at a specific distance from the nucleus.

The above is only the most cursory examination of atomic structure and only a small amount of the information which can be inferred from the diagram under consideration. Despite this, it will serve to offer a very logical and hard-to-disprove explanation of light emission using the following argument.

(a)   An atom can be forced to accept extra energy into its system. A common example of this is applying heat energy to a bar of iron and causing the temperature to rise.

(b)   The most usual way an atom accepts extra energy is for one or more electron of the structure to absorb a discrete portion of the total energy and move into a higher (and unoccupied) energy band. This is shown in figure 7/24.

(c)   With an electron or electrons at a higher than normal energy level, the atomic structure is in an unstable condition. Sooner or later the electron or electrons will fall back to their original, now only partly occupied level. This original level is called the 'ground state'.

(d)   In returning to the ground state from a higher energy level, the electron must shed some energy. This it does in a minute packet or train of electro-magnet waves called a 'photon'.

(e)   If the energy emitted is high (that is if the jump back to the ground state is large) the photon will carry a high frequency, well up the electro-magnetic spectrum, see figure 7/4. If the energy is lower the frequency is lower and hence this frequency is shown further down the spectrum. High energy photons in that portion of the spectrum of interest here are ultra-violet photons. Lower energy photons will be in the visible part of the spectrum. A photon of violet light carries twice the energy of that in a photon with a frequency giving red light. Lower energy again will place the photon as an infra-red (heat) photon. Figure 7/4 therefore becomes

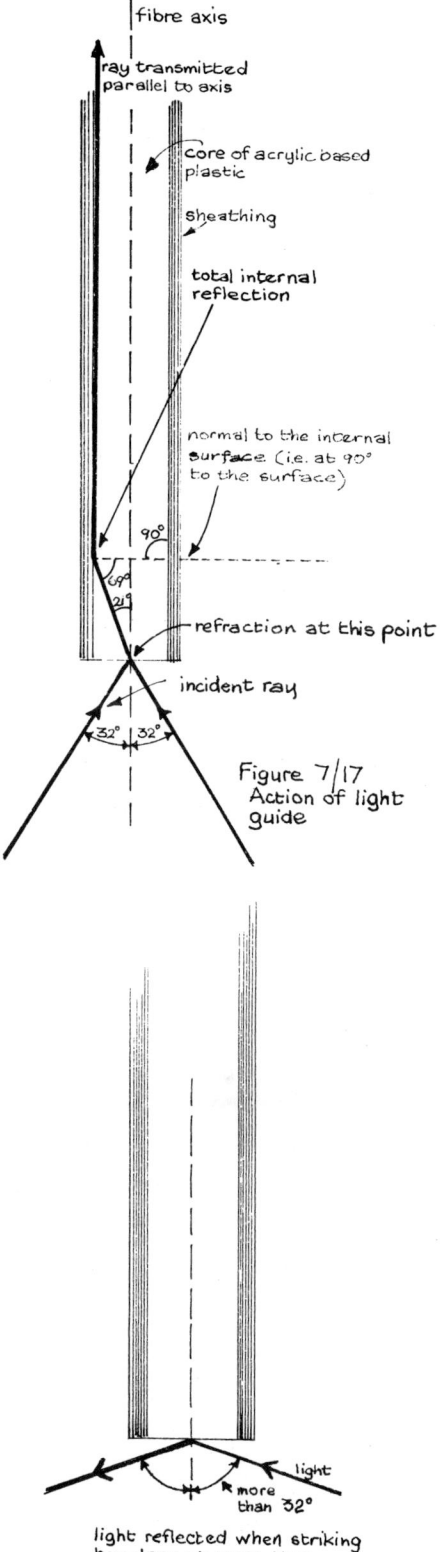

Figure 7/17
Action of light guide

light reflected when striking boundary at more than critical angle

Figure 7/18

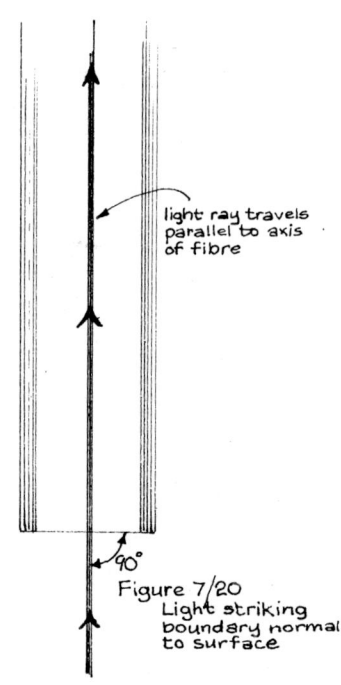

$90° - (x - 11)°$

$90° - (x - 11)°$

$(x - 11)°$

where angle x is less than 32°

$x°$

Figure 7/19
Light striking
boundary at
less than critical
angle

light ray travels
parallel to axis
of fibre

90°

Figure 7/20
Light striking
boundary normal
to surface

more meaningful as an indicator of the relative energies locked up in photons of different frequencies.

Heat is not the only form of energy that atomic structure will accept. Some very special conglomerates of atom will accept the energy of moving electrons which we know as an electric current. These, as yet very rare substances, have electrons associated with the atoms making up the substance, which relatively easily move above the ground state when the material is connected to the normal electricity supply. The energy jumps of the electrons in these materials are of such a value as to emit photons of such a frequency as to come into the visible range. The general term for the phenomenon which can be caused to emit light in this way is electro-luminescence.

When not glowing with light the most common, commercially available luminescent material looks like shiny plastic. It can be obtained in sheets, plates, rods and filaments. When connected to the mains it glows a soft, low-intensity light. The scope for such as backgrounds for visual compositions, as flexible fibres threaded around and through designs of various sorts, cut to shapes, bent to curves, etc., is most exciting. Electro-luminescent substances can be strongly recommended. They are capable of some spectacular applications and are sufficiently new to have the appeal of novelty as well.

Light emissive diodes (LED)
One of the latest developments in the opto-electronics field is a tiny light source, an offshoot of the transistor industry. There is little that is spectacular about a LED apart from its small size, low power consumption, low voltage (approx. 3 volts), long life, ruggedness and comparatively small cost. Developmental work on LED's is continuing but as at present the colours offered are limited and the actual area of illumination is very small so the LED is often visible only head-on.

The principle of operation of the LED is that of the so-called tunnel diode as used in pure electronics. The actual theory is not well understood and what is generally accepted is complex. However, it is reasonable to describe photon production in a LED as being the result of electron energy in the atoms of the device being raised by the action of an electric current flowing across the barrier gradient between the two halves of the diode. Very roughly this can be considered to be akin to water across the water-fall carrying cork floats under the surface to bob up with sufficient energy to leap clear of the water surface.

It must only be a matter of time before some enterprising visual artist uses the small size and low power consumption of LED's as the basis of an illuminated display which can quite well give pseudo-movement by pro-grammed switching of banks of LED's. It is not proposed to pursue this matter here other than to suggest that rapid developments in this field make it one of continuing interest to the serious contemporary artist or craftsman.

Fluorescence
A very wide range of minerals and chemical substances have the atoms of their structure locked in such a way that high-energy, ultra-violet photons can place the orbital electrons of the material in an above ground state, energy level. When these excited orbital electrons fall back to ground state they emit photons in the visible spectrum frequencies. This is the basic mechanics of the phenomenon of fluorescence.

By choosing the correct 'phosphor' (as materials which exhibit fluores-cence are called) a wide choice of colours are offered. The human eye is not responsive to ultra-violet radiation (showers of ultra-violet photons) although some parts of the eye fluoresce in strong ultra-violet, causing space to be filled with an unpleasant milky-blue radiance. The absence of reaction of the eye to ultra-violet allows fluorescent materials to be self-illuminated in otherwise near darkness. Judiciously used, this property of fluorescence can allow some unusual effects to be produced.

Sunlight is rich in ultra-violet, fluorescent lamps rely on the principle, the light source being actually a layer of fluorescent power inside the glass of the envelope of the lamp and ordinary incandescent lamps emit small quantities. The normal visible light from each of these sources swamps the fluorescent illumination although an added depth of colour can be given to certain coloured pigments in this way.

The most satisfactory source of ultra-violet for exciting fluorescent materials is one or other of the commercial, mercury vapour, discharge lamps with quartz arc tubes. An electric discharge (arc) in an atmosphere of mercury vapour is a rich source of ultra-violet radiation. This is the primary source in the common fluorescent tube lamp. Ordinary glass is opaque to ultra-violet and quartz is used instead of glass as the ultra-violet window in the special lamps.

Two general classes of ultra-violet light lamps are offered. One the so-called 'long wave UV or blacklight' lamp which produces photons with frequencies predominately in the 3650 Agström band. Little visible light is emitted from these lamps as an outer bulb of Wood's glass which is opaque to visible light but nearly transparent to ultra-violet surrounds the quartz tube. These lamps need special control gear but can be recommended for exciting the more common fluorescent substances. The second general class includes the so-called 'short-wave UV' lamp which emits over a far wider spectrum than the first, including the visible band and is not recommended for the applications being discussed here.

Acrylic based fluorescent colour media are freely available from stockists of artists' supplies. These paints must be used with considerable discretion but have a part to play in ordinary modes of painting. Fluorescent red over a red ground colour gives a fiery appearance not otherwise, achieved. Light washes of fluorescent yellow over pastel shadings can lend brilliance and produce highlights. Overdone, the fluorescent colours can dominate and ruin a composition, although when used for sheer impact value they are singularly successful. All these applications are for viewing in ordinary light the special characteristic of fluorescence being used to supplement other effects.

To obtain the maximum shock value, fluorescent materials and paints need viewing under very subdued lighting or total darkness, fluorescence being produced by a blacklamp as described earlier. Some complications will probably arise. A surprising number of present day articles are made to appear white or in their true colour by blue fluorescent dye (sometimes other colours) being incorporated in the material. Fabrics and paper are two substances prone to contain dye which cause them to glow bright blue with self-illumination under ultra-violet. The blue colour is introduced to counter the natural yellowing of these substances. Printed fabrics, in greens, yellows and reds are often highly fluorescent and can be the starting point of novel fluorescing collages.

The aforementioned special paints can be used to great advantage under ultra-violet but many papers must have their fluorescent properties killed with a heavy coat of non-fluorescent opaque paint before being suitable. Many minerals will fluoresce. A partial list is given now, the figures refer to the wavelength which excites maximum fluorescence.

Angelsite 3650 Agströms (A)	Scheelite 2900 A	Fluorite 3200
Dolomite 2750	Willemite 3350 A	Kunzite 3700 A
Gypsium 3650	Calcite 3120	Sphalerite 3650 A

It must be noted that the figures given can suffer some variability as between different specimens of a given mineral and indeed some specimens do not fluoresce at all. There are literally hundreds of other examples of fluorescent substances but the ones suggested can be found in massive deposits, at times large enough to sculpture. It should not be overlooked that small pieces of fluorescent materials can be embedded in a matrix of

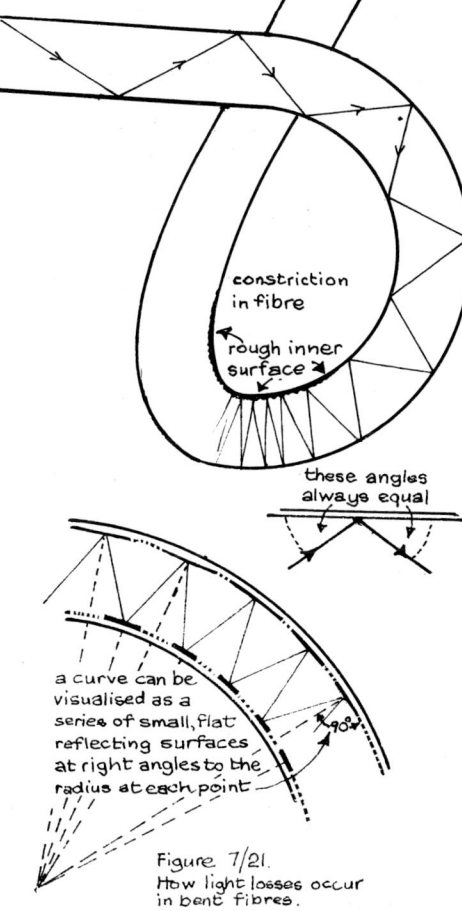

constriction in fibre

rough inner surface

these angles always equal

a curve can be visualised as a series of small, flat reflecting surfaces at right angles to the radius at each point

Figure 7/21.
How light losses occur in bent fibres.

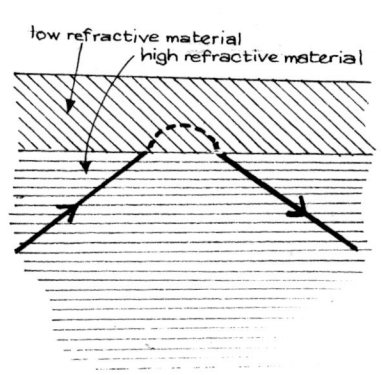

low refractive material
high refractive material

Figure 7/22
Reflection of light

plaster, cement, plastic or other sculptural media or arranged into panel and other designs.

Liquid crystals

To folk with a passing acquaintance with physical science the expression 'liquid crystal' seems a contradiction in terms. It is at that, but it is descriptive of a class of liquid substances which show some properties of a liquid and some properties which are usually associated with such crystalline materials as mica, quartz and diamonds. In particular liquid crystals exhibit many of the optical characteristics of solid crystals with some fascinating peculiarities of their own.

A sandwich can be made with a layer of liquid crystal between two glass plates, one of which is made reflective like a mirror and both are coated with an electrically conductive but transparent coating. When an electric potential is established between the two plates the liquid crystal can be made to go frosty and so appear as frosted glass with the mirror image lost. Removing the potential can cause the liquid to return to its normal clear state. This is only one of the many quite astonishing effects which can result from altering the normal crystalline structure of these peculiar liquids by applying electrical charges.

A conducting mosaic of numerous separate elements can be laid down on a glass sheet and each element brought out to separate leads. A second sheet is either purchased as conducting glass or coated with a transparent tin oxide layer to make a conducting surface. A layer of liquid crystal is placed between the sheets and sealed with epoxy resin binding the edges of the sheets. By applying a voltage between the elements of the mosaic and the tin oxide conducting layer the liquid can be caused to change its optical qualities. With certain substances colour changes can be instituted at different values of voltage. The leads brought out separately from each element allow them to be connected in various combinations to form patterns which may be changed at will.

There are well over 200 organic compounds known to exhibit liquid crystal characteristics, many of them falling into what is known as the cholesteric groups—cholesterylchloride, cholesteryl nonanoate and cholesteryl oleate being representative examples. These compounds and others, together with electronic conducting glass, are reasonably easily obtainable from science supply houses. The voltages required are not high; from 3 volts to 50V or so of direct current being required in accordance with the arrangement of the apparatus. There is not a great deal of literature on the subject outside of very specialised papers, but some information is being given from time to time in popular science magazines.

In addition to the electro-optical changes of liquid crystals, a good number are extremely temperature sensitive, showing a distinct change in colour for as little as a 3°C change in temperature. Most polorize light and this property is pressure sensitive so that colour patterns are exhibited when a liquid crystal is compressed and viewed through a polarizing screen. Altogether, these materials are as fascinating a range of experimental potentials as any under development today. There is tremendous scope for entirely original and attractive optical displays in this field.

It is good to have touched upon some of the latest developments in the craft of lighting. This chapter may serve to show the exciting possibilities opening in the general area of decorative and artistic lighting. For artists with the will to develop creatively in the fields of technology and for the technologists with a willingness to give reign to their artistic feelings there is a wealth of satisfaction in the use of light in its many forms. Artistic lighting is a mode of expression as yet little enough exploited. It is hoped that this book has gone at least part of the way towards exploring ways in which truly appealing lighting can become more than just a means of providing illumination.

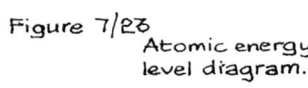

radius of influence of orbiting electrons

electron energy levels

N shell

M shell

L shell

K shell

atomic nucleus

Figure 7/23
Atomic energy level diagram.

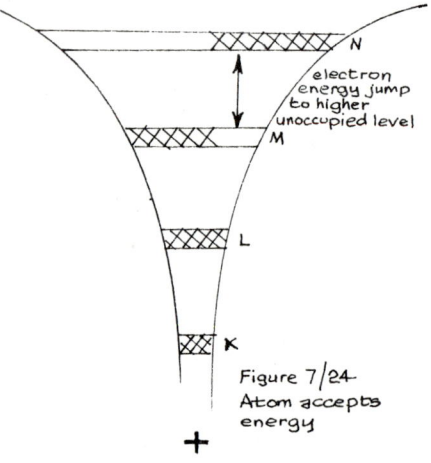

electron energy jump to higher unoccupied level

N

M

L

K

Figure 7/24
Atom accepts energy